MINING and the ENVIRONMENT

The Berlin Guidelines

A study based on an International Round Table in June 1991, organized by the Department of Technical Co-operation for Development, United Nations, and the Development Policy Forum of the German Foundation for International Development.

Mining Journal Books

Mining and the Environment

First published 1992. Reprinted 1995.
Mining Journal Books Ltd, 60 Worship Street,
London, EC2A 2HD, U.K.

ISBN: 0 900117 56 7

DOK 1634 B/a
IT 04-02-91

Text set in 11/13 pt Times Medium Roman by photo-composition and printed offset litho by Chandlers Printers Ltd, Bexhill-on-Sea, East Sussex, England.

Contents

APPENDICES

Foreword

At the end of June 1991 an international Round Table conference entitled 'Mining and Environment' was held in Berlin. The meeting was organized by the Development Policy Forum of the German Foundation for International Development (DSE) and the United Nations Department of Technical Co-operation for Development (UN/DTCD). As of March 1992, the functions and programs of the UN/DTCD are being carried out by the newly formed United Nations Department of Economic and Social Development.

Although conferences on the environment are now proliferating, this particular meeting was more important than most for two reasons. Firstly, it was organized by institutions with influence on the availability of funds. Secondly, the meeting addressed the central issue by concentrating upon the long-term sustainable development of the mining industry in countries with different levels of industrialization. After four days of discussion, the meeting reached a compromise on a set of environmental guidelines (which are reproduced as Chapter 10).

By convening the Round Table on Mining and Environment, the UN/DTCD sought to address issues related to the role of mineral resources in achieving sustainable development. The main purpose of the meeting was to formulate international principles, guidelines and national regulatory standards regarding mining, with particular emphasis on developing countries in order to initiate a consultative process between the various actors involved, such as the mining industry and developing country governments.

A need for international guidelines has been recognized as vital by international mining companies and international development agencies. The meeting discussed the needs and mechanisms for better co-ordination and co-operation among, and between, bilateral and multilateral development agencies in respect of mining and environment activities in developing countries. Global problems, economic and non-economic alike, provide ample evidence of the need for more effective international development co-operation.

The United Nations system of organizations has been at the forefront of efforts to place environmental problems on the world's agenda. In response to the first great wave of public attention to environmental issues in the 'West' during the late 1960s and early 1970s, the UN held its first Conference on Human Environment in Stockholm in 1972. This resulted in the Stockholm Declaration on Human Environment, which initiated a number of international conventions regulating global environmental problems. Another important result of the conference was the establishment of the United Nations Environment Programme (UNEP), the main UN body mandated to promote sound environmental practices.

The environmental issues continued to evolve during the 1980s. The report of the World Commission on Environment and Development – also known as the Brundtland Commission – Our Common Future, raised a call for decisive political action to link economic growth and development with sound management of our environment. The Brundtland Commission report underlies the crucial role of the UN in translating the rhetoric of sustainable development into operational standards against which state policies can be evaluated.

Increased networking and interaction is taking place between the UN and other intergovernmental organizations and development banks, as well as among non-government organizations, research institutions and experts. The World Bank, the UNDP and UNEP have established the Global Environment Facility with $US1·3 billion of expected funding to provide grants and low interest loans to countries to ease environmental problems. For the first time the private sector enterprises have also become active constituents of these emerging networks. An important initiative was launched in April 1991, when the International Council on Metals and Environment, an organization of mining companies, was formed to promote sound environmental policies and practices. Its priorities include developing an environmental stewardship policy and working within the UN to create environmental guidelines for developing countries.

The UN/DTCD's agenda is broad and encompasses a number of issues. These include:

- economic, social and political processes influencing the development of regulation and environmental management in the mining industries of different countries;
- economic, social and political factors determining the behaviour of national and international mining companies towards the environment and the implications of corporate responses to increasing environmental regulation for patterns of resource use, investment, metal markets, technological innovation and population and labour, including the role of women;
- the role national governments, industry and international agencies play in developing and diffusing new technologies that reduce pollution.

It is the hope of the UN/DTCD and DSE that the Berlin Round Table will have contributed towards formulating the necessary environmental guidelines and strategies on mining with emphasis on developing countries.

We are all aware that action is not only urgently needed but also long overdue, and we have an unprecedented opportunity to take that action. We owe it to our children and grandchildren to leave them a planet for life and capable of sustaining its future.

World Summit

More than twenty years after the environment issue was first placed on the UN agenda, the World Summit Conference devoted to Environment and Development will be held in Rio de Janeiro, Brazil, from 3-14 June 1992. Much of the preparatory work for this conference is focusing on issues such as the protection of the atmosphere, protection of land resources and fresh water resources, protection of oceans, seas and coastal areas, protection of biological diversity, biotechnology, management of hazardous wastes and prevention of illegal traffic in toxic products and wastes. The interdisciplinarity of the issues involved, the universality of their scope, and the search for action-oriented solutions involving financing, transfer of technology, institutional and human development, and related legal, social, cultural and political implications pose a major challenge to the international community.

The conference in Brazil will result in the *Earth Charter* which should embody basic principles to guide the economic and environmental behaviour of nations and peoples. *Agenda 21,* another expected outcome of the conference, will comprise a blueprint for action for the twenty-first century. According to Maurice Strong, the Secretary-General of the Conference, it will not be a static plan of actions, but a dynamic instrument through which the governments and all other participants can manage and monitor the process.

The issues of environment and development have taken an increasingly important place on the international agenda since the Stockholm Conference in 1972. Whereas earlier environmental degradation was seen largely as a problem of industrialized nations, it has in fact become recognized as a survival issue for developing countries today, in view of the need for their intensive growth and development. At the present time, both 'North' and 'South' tend to agree on the importance of safeguarding the natural environment and on the necessity to achieve sustainable development, ie development which meets the needs of the present without compromising the ability of future generations to meet their own needs.

Dunja Pastizzi-Ferencic
Director
Natural Resources and Energy Division
UN/DTCD
New York

Dr Götz Link
Executive Director
Development Policy Forum
DSE
Berlin

Acknowledgements

This volume contains a selection of the papers presented at the Berlin Round Table of June 1991. This meeting, entitled Mining and Environment, was organized by the United Nations Department of Technical Co-operation for Development and by the Development Policy Forum of the German Foundation for International Development.

Because of budgetary limitations, and the need to present a concise summary of the proceedings, even the presentations selected for publication have been edited (in some cases drastically). However, considerable effort has been made not to modify the substance of the original papers or the opinions of the authors. In this regard the authors are thanked for their forebearance in accepting the necessary changes.

The organizers and participants are grateful to many individuals concerned with the meeting, and this subsequent publication, but special mention must be made of Heino E. Wittrin, Chairman of the Meeting in Berlin, and of Beatrice Labonne and Antonia Cubeiro of the UN/DTCD in New York.

CHAPTER ONE

Introduction to the Issues

Edited extracts from introductory statements made at the Berlin Round Table by Dunja Pastizzi-Ferencic, Director, Natural Resources and Energy, and Beatrice Labonne, Chief, Mineral Resources Branch. Both from the UN/DTCD in New York.

The most outstanding issue still to be resolved is to determine the nature of the linkages between environment and development, and how to lay the foundation for a global partnership between developing and industrially developed countries, based on a recognition of mutual needs and common interests.

There is a general consensus that three sets of linkages should be addressed. One of the most important linkages tieing environment and development is that between poverty, demographic pressures and development. Another set of linkages between environment and development now being explored is that between growth patterns, consumption standards and environmental consequences. The third set of linkages identified relates to the international dimension of development, such as debt and adjustment, deterioration in terms of trade, and reverse financial resource flows.

By emphasizing these linkages between environment and development, the world conference in Brazil will address the well-known reality: that prescriptions for environmental action by themselves, without resolving major developmental problems of our times, simply cannot work.

In the current environmental debate there are certain crucial 'blind spots', such as the environmental impact of mineral development. The linkages between development, mining and environment are to be still further explored and need to be brought on to the international agenda for action.

The impact of growing population and economic activities has already weakened the natural resource base of many countries, but the ability to fulfil the needs of present and future generations increases with scientific knowledge and the development of ecologically sound technologies. It is generally recognized that the problems of poverty – a major environmental degradation – cannot be solved without vigorous economic growth. The concept of developmental and environmental sustainability will require changes in current patterns of growth to make them less resource and energy intensive and more equitable, taking into account the requirements of the growing population.

Although the mining industry is not a large employer in industrialized countries, the situation can be different in developing countries where mining often represents one of the few opportunities for formal employment. Mining also offers linkages into other parts of the economy, and the indirect employment effects of mining in developing countries are considerable. Moreover, small-scale mining has been identified as a vehicle for the creation of an indigenous entrepreneurship which is a vital force in the process of industrial development.

An important set of linkages deals with trade-related issues. The value of world trade in minerals is considerable. While world exports of metal ores and scrap have not regained their 1980 figure of \$31·8 billion, they reached \$27·1 billion in 1987 from a low point of \$24·3 billion in 1986. Moreover, the share of exports accounted for by developing countries has remained at about one-third throughout this period, which means that developing countries depend heavily on export earnings from the mining sector.

The concept of sustainability in relation to mining requires recognition that the mining sector is of crucial importance to developing countries. However, mining activities are generating considerable environmental impact on the physical environment such as pollution of rivers, deposit of tailings, destruction of the natural landscape, air pollution by dust and gases, disruption of marine environment in case of off-shore mining as well as on the socio-cultural environment affecting and destroying social patterns of local populations. That is why it is imperative to assess first the entire ecosystem and general developmental context which may be affected by present and planned mining operations. However, at present in most developing countries, the mining industry is often given high priority. Measures to manage environmental and socio-cultural impacts are considered a luxury more fit for the rich 'Western' countries than for developing countries suffering from well-known financial constraints.

> Sustainable development is about meeting the needs of the present without compromising the future.

The current legal discussions point out that the right to development and the concept of permanent sovereignty over natural resources ought to be compatible with the concept of developmental sustainability, ie sustainability in relation to resources available for present and future generations, sustainability in relation to an adequate and healthy environment, sustainability in relation to local, national, regional and international communities.

1.1 Mining: A Dirty Word

Unfortunately for the environment-conscious miner, ore deposits must be developed *in situ,* and the extractive companies must survive in a competitive and volatile world market. Furthermore, the mining industry differs from other productive industries in generating an incredibly large amount of solid, gaseous and liquid waste material mainly in the form of overburden, tailings heaps, slag, sludge, mineralized toxic waters, smoke and gas emissions. Moreover, mines are all different from each other.

Environmental protection is a relatively new concern. It is still difficult to obtain reliable statistics on the cost of environmental protection in the mining sector, and the sparse information available primarily refers to the industrialized countries. Nevertheless, it is clear that pollution as a result of mining and smelting activities is significant in comparison with other industries, and that the mining industry recognizes its responsibility with regard to protecting the ecosystem. But how much will it cost and who will pay for it?

People can choose between environmental luxuries and environmental necessities. One must tend towards achieving a manageable and acceptable level of pollution rather than striving to reach the zero pollution level. A modulated approach is recommended, one that can balance economics, ie growth against health concerns and our social obligations towards the future generations. With regard to an acceptable level of pollution, one may ask if some countries are more able to support a higher degree of pollution than others?

1.2 Free Riders

What will be the development strategy of mining investors as a result of tighter environmental legislation by the governments of industrialized countries? Recently, both Australian and Irish media reported that mining investors will be forced to look overseas to avoid becoming entangled in domestic legislation that is either too unclear or stringent for their liking. At the outset, in the developing countries the options are limited for the key players, ie the mining companies in need of mineral commodities, and governments in need of revenue but committed to the social well-being of their people.

Is the cost of environmental protection prohibitive for the developing nations? What are the trade-offs that developing countries must make between the costs of preservation of the environment and the benefits of a mining project? And how can these costs be estimated when social and economic growth is at stake? Is it even advantageous for a government to compromise its pledge to achieve sustainable development in favour of short-term gains? Is there a balanced approach? A positive development alternative? Will the developing countries be assailed by scores of 'free riders' on the prowl for lenient legislation which will help to reduce their investment costs and give them a competitive edge?

Environment-related costs are difficult to isolate, since in the case of new project developments they tend to be incorporated in the capital and production costs. Only those costs resulting from rehabilitation of existing mines and smelting plants, or clean-up operations after mine closure, are 'add on' expenditures and thus readily apparent.

Darwin's law also applies to the extractive industry: unless it adapts it is bound to become extinct. Only those companies which modernize are likely to survive. To avoid unbearable increases in production costs, or closure, resulting from environment restrictions, mining companies must choose an environment-friendly technology which will also have to contribute either to increased production or improved productivity.

Mining technology, environment technology inclusive, is often wrongly perceived to originate in industrialized nations because they are competing in a market economy, and enjoy an economic advantage. However, mining technology is developed where there is a strong and dynamic mining industry supported by a solid economic and financial base. It is therefore possible that large mining countries, like Brazil, are able to develop a world-class technology that is well adapted to the needs of the developing countries, and which can be readily exported to the international market. Such state-of-the-art technology may be more affordable and better suited for the developing countries which often face the same environment and climatic conditions.

Many developing countries, particularly in Africa, with a good geologic potential are worried that their efforts towards improving their investment climate might be in vain – that the investors will abandon them for Eastern Europe which has an abundance of good mineral deposits. But mindful of the saying 'new mines are often found near old ones', that region will have to be rehabilitated before any new mining venture takes place. This may place otherwise valuable deposits in an unattractive economic position compared with an African orebody lying in the wilderness.

In the context of costly environment protection, what are the survival chances of the small mining operations or even those of the marginally economic projects which cannot support a costly, sophisticated technology? The medium/small-scale mining industry may end up being the main casualty of the enforcement of environmental protection legislation. These companies are typically owned by local entrepreneurs, have limited capital resources, frequently suffer from poor management, utilize primitive technology, and pay low salaries to unskilled workers who often operate on a part-time basis. Too often these companies tend not to be real revenue generators and contribute little to government income. However, they benefit the community by providing or supplementing local income. Unlike artisanal mining, which in most instances operates in a clandestine fashion, small-scale mining will suffer economically, unable to meet the cost of the 'green' mining legislation.

Who is going to pay, and how, for 'upgrading' the mining industry, ie the costs of impact studies, control technology, rehabilitation, and decommissioning and land reclamation? The tax payer and/or the consumer seem to be obvious victims. One way of alleviating the burden on both the environment and the consumer is to stimulate a meaningful mineral conservation strategy. This is a strong medicine for the industrialized countries and likely a lethal one for the developing nations. The cost of conservation technology is a complement to that of environment protection. In its 1991 report, the Intergovernmental Panel on Climate Change sought to limit the level of gas emissions responsible for the greenhouse effect, and suggested a flexible and phased response which relied on a mix of environment protection and conservation measures. Similar strategies can be applied to the mining industry.

1.3 Incentive and Tax

Environmental objectives can be pursued through a complex mix of regulations, market-based economic instruments and financial incentives. Each government will have to strike a balance between incentives and taxes to fill its coffers, while preserving the eco-system. Many governments of developing countries will opt to set up an array of administrative and legal regulations to enforce their new environment guidelines. However, experience indicates that many developing countries have already emplaced relevant and realistic policies but have been unable to enforce them. In the developed countries, economic instruments and other financial and tax incentives have, through their flexibility, encouraged innovation and contributed to the development of new and performing technologies. The resulting environment improvements were achieved at a lower cost for the industry than through the sole regulatory mechanisms.

The 'polluter pays' principle can also be simultaneously applied to ensure that the stubborn companies comply. We may smile hearing that a US company was fined $25,000 each time a visiting duck was found dead as a result of cyanide poisoning, but the unfortunate ducks served the purpose and subsequently the company was able to solve its pollution problem. What are the options available to the developing countries, whose priority is sustainable development, and who can therefore ill afford to destroy any revenue-generating industry? A gradual enforcement of regulation may be the best solution.

Since the mining industry will have to clean up its act literally, financing remains the main concern. Companies operating in developing countries are placed in a more disadvantageous situation, as these countries do not have the financial resources to fund their environmental programmes. External resources will have to be identified. As already indicated, tax and other incentives in conjunction with progressive legislation could play a major role in attracting private investments for project development. There are a number of different approaches to ensure that a project is not delayed because of high environment-related costs. One possibility is for these costs (within defined limits) to be tax deductible; this applies specifically to funds set aside for reclamation.

In addition to the usual financial incentive tools, debt-for-nature-swaps may be envisaged. Some countries have already promoted them. For all the hard currency debt involved in the swap, a country could undertake to set up, and develop, relevant environmental programmes financed in local currency. Additional financial resources to support these protection activities may be in the form of grants and loans at concessionary rates.

Finally, an awareness of the need for protection of the environment must also be instilled, most particularly in the developing countries, where few have access to satisfactory levels of education. Training is an element to take into account when addressing the cost of environment protection.

CHAPTER TWO

Environmental Regulations

Edited extracts from papers presented by Dr Lawrence J. MacDonnell, Natural Resources Law Center, University of Colorado School of Law, and by Dr Alyson Warhurst, Science Policy Research Unit, University of Sussex.

2.1 Regulations in Selected Industrialized Countries

Although there is considerable variation in the degree of environmental regulation of mining, even amongst leading mining producers in the industrialized world, there is also a surprising amount of commonality in the general approaches that have been taken. The common elements in environmental regulations which apply in the U.S., Canada and Australia are briefly summarized here.

1. Land areas determined to have special environmental values have been specifically reserved from mining activities. These areas commonly include national parks and wildlife management areas. In addition, processes have been established to consider environmental losses that may result from proposed new mining activity elsewhere. Generally there is discretion to preclude mining in these areas if the environmental losses are considered unacceptable.

2. An environmental impact assessment process is utilized to inform governmental decision-making of the environmental consequences of mining activities requiring governmental approval. The assessments are not themselves decision documents but rather

are intended to ensure that full consideration is given to the environmental impacts of the decisions. They may result in government approval being conditional upon the incorporation of mitigation measures or in the identification of more environmentally acceptable alternatives.

3. The pollution-generating aspects of mining activities are subjected to permitting requirements which limit the discharges of wastes according to specific performance requirements. Discharges to surface water are comprehensively regulated. Management of wastes, especially in relation to indirect effects on surface and groundwater, is gaining increased attention. Monitoring requirements are commonly a part of permit programs. Violation of permit conditions results in various enforcement options.

4. Reclamation of the surface area commonly is required upon cessation of mining operations. The nature of the reclamation requirements varies widely but typically includes revegetation and protection of surface water resources. The posting of a bond as security for performance of reclamation is increasingly common.

2.1.1 Environmental regulations in leading mining countries

United States

The environmental effects of mining are subject to a wide array of laws and regulations in the U.S. Virtually every aspect of mining, from exploration through reclamation, comes under some form of either federal, state, or local government control. A brief summary is provided here.

The use of land for mining is no longer automatically considered to be its highest use. Large areas of the public lands (those lands, comprising approximately 30% of the land area of the U.S., owned by the federal government) are not available for mineral exploration and development. These include national parks, military reservations, and wilderness and wildlife protection areas. Mineral development on other parts of the public lands are subject to control by the federal agency with management responsibility for these lands. For example, exploration for hard rock minerals on national forest lands involving significant disturbance of surface resources requires a plan of operation to be approved by the U.S. Forest Service. State and local law may also affect mineral exploration and development activities on private lands. For example, the Colorado Mined Land Reclamation Act requires exploration and development activities to be conducted in a manner that will minimize surface disturbance. Thus a trend in the U.S. is simply to preclude mining activities on certain lands because of their value for environmental purposes and to require minimal surface disturbance on other lands during the exploration process.

Mining represents an often irretrievable commitment of the associated land area. Increasingly, this commitment is being scrutinized because of concerns about environmental losses that might result. The U.S. has enacted a series of statutes aimed at protecting land, water, air and wildlife. Activities threatening the related human health and ecological concerns are subjected to permitting requirements aimed at requiring specific performance standards determined to be acceptable. The laws are not directed at comprehensive regulation of different types of activities (such as mining). Rather, they aim at protecting things like ambient air or water quality. To construct and operate a mine in the U.S., the company will have to obtain a series of permits from federal and state agencies. A few examples are provided next.

The Clean Air Act limits emission of pollutants into the atmosphere to protect public health and welfare. Permit conditions apply primarily to mineral processing activities but may also apply to control the particulates of dust that are generated during mining. The Clean Water Act subjects all discharges into waterways to a permit requirement. Pollutant concentrations in these discharges are specifically limited according to certain technological treatment requirements. The Resource Conservation and Recovery Act requires that all solid and hazardous wastes be managed in an environmentally safe manner. A regulatory program for mining wastes is in the process of being established.

The National Environmental Policy Act subjects major federal actions significantly affecting the quality of the human environment to the requirement of an environmental impact statement (EIS). Permitting decisions of the kind described above may trigger the need for such an EIS. The purpose is to ensure that federal decisions are made with a full understanding of the probable environmental consequences. In practice, environmental problems identified in the EIS are likely to be addressed through conditions attached to the permits.

The permits essentially define the acceptable level of pollution creation and other environmental impacts that may accompany the mining activity. In particular, they define the degree and manner in which these pollutants may be released from the facilities into the air, water and land. In addition, they place ongoing monitoring and reporting requirements on the operator. Failure to operate in compliance with permit requirements may result in substantial fines, curtailment of operations, and even criminal penalties in some cases.

Reclamation of the land surface is subject to federal control on public lands and state control on private lands. Concern about the effects of surface coal mining led to the enactment of the Surface Mining Control and Reclamation Act which generally requires restoration of the original surface.

Australia

Unlike the U.S., mining in Australia is controlled largely by state law rather than at the national or Commonwealth level. All mineral resources in Australia are owned by the Crown. Legal regimes providing access to explore for and develop mineral resources are established by the states. In general, the state issues exploration licenses and mining leases. Considerable discretion is vested in the government in the award of these titles.

The system of environmental regulation is not as comprehensive in Australia as in the U.S., but it has developed considerably in recent years. As in the U.S., substantial areas of land in Australia have been closed off to mining or allow mining only on a restricted basis. In 1988, the Australian Mining Industry Council stated that 23% of the total land area in that country is not available for mineral exploration or development. This includes parks and nature preserves but also areas protected for social and cultural reasons (Aboriginal and pastoralist areas).

The mining laws of most Australian states provide that environmental factors should be considered in issuing exploration or production leases. Thus the New South Wales Mining Act directs consideration of the need to conserve and protect flora and fauna, as well as scenic and archeological values in the area to be leased.

The Commonwealth and all the states have environmental impact assessment requirements. The objective of the Commonwealth law, enacted in 1974, is to ensure that matters affecting the environment to a significant extent are fully examined and taken into account in the actions of the government. The initial decision regarding the need for an assessment rests with the Minister responsible for the proposed development (the 'Action' Minister). The matter is then referred to the Department of Environment for a final determination. If an assessment is deemed to be necessary, the Action Minister consults with the Environment Minister in its development. Because control of mineral development rights rests with the states in Australia, environmental assessments of mining projects are most likely to occur under state requirements. As mentioned, several state mining laws provide for review of mineral leasing decisions on environmental protection grounds.

Pollution control in Australia also is the subject of state regulation. As in the U.S., the general pattern is to establish separate statutes for air pollution, water pollution and waste disposal. Discharges of pollutants can occur only under a government-issued license that establishes minimum performance requirements. The licensing authorities are given broad discretion in establishing license conditions.

State mining laws generally require rehabilitation of the land surface following mining activities. For example, in New South Wales mining leases typically contain conditions requiring rehabilitation which may include revegetation, contouring and other actions. A bond ensuring that the rehabilitation measures will be carried out is commonly required. South Australia has established an 'Extractive Areas Rehabilitation Fund' supported by royalties from mining and used to cover costs of rehabilitation and to fund research into improved mining techniques that will reduce environmental damage.

Canada

Mineral ownership patterns in Canada are similar to those in the U.S., in that lands passing into private ownership in the 19th century generally carried mineral rights with them. Subsequently, the mineral rights stayed with the Crown. The federal government controls mineral development in the Northwest Territories and the Yukon, while provincial governments control Crown rights within their provinces. Therefore, regulation of mineral exploration and development varies among the jurisdictions in Canada.

Typically there is very limited control of exploration activities. Prospecting licenses generally are required, and federal and provincial laws have excluded land areas such as parks, game preserves, and bird sanctuaries from prospecting. In addition, the minister with authority over Crown minerals has broad authority to withdraw additional lands from mineral activities.

Substantially greater scrutiny is given to proposed development and production activities. Typically a production lease is required for such activities. A screening process is used to make this decision which involves consideration of environmental effects and development of mitigation measures. In general, the process is administrative rather than statutory. It is a consultative and interactive approach involving informal meetings between the mining company and the government representatives. There tends to be little public involvement in the process.

Each province has enacted statutes subjecting discharges to air and water to a permitting system. The general approach is similar to that in Australia and the U.S. The federal Fisheries Act has established a nationwide program of regulation for the protection of water. Under this program, regulations specifying quantitative levels of dissolved metals, pH, and suspended solids that may be discharged from a mining property have been established.

Reclamation requirements vary by province as well. Alberta has established a detailed reclamation program for coal mining which requires the filing of a development and reclamation plan including reports on surface disturbance, surface and ground water management and pollution control. British Columbia requires a program for protection and reclamation of the land surface and water course that will be implemented upon completion of mining activities. Approval of the program is entirely discretionary with the Minister of Energy, Mines and Petroleum Resources. Security for performance of reclamation is generally required.

Broad Concensus

This summary suggests a broad consensus among the U.S., Australia and Canada on policies regarding mining and the environment.

This consensus on the common approaches will need to be modified, or adapted, to meet local conditions. In addition, the concept of an international convention on mining and the environment should be considered.

Specific questions for debate include:

1. Should specific land areas be excluded from any mining-related activities because of their natural, environmental and heritage values?
2. Should World Heritage listings be used for this program?
3. Should a governmental agency have the discretion to exclude additional areas proposed for mining activities on environmental grounds?
4. Should the environmental effects of mining activities be subjected to an assessment process?
5. Should the 'Convention on Environmental Impact Assessment in a Transboundary Context' be extended to all mining?
6. Should the regulatory permit system be applied to pollution-generating activities of mining?
7. Should the 'UNEP Environmental Guidelines: Restoration and Rehabilitation of Land and Soils After Mining Activities' be extended to all mining?

8. Should a Task Force be created to develop a manual of 'best industry practices' relating to mining and the environment?

2.1.2 Environmental regulations in primarily smelting/refining countries

Japan

Japan has a small land area and few mineral resources to support its population of over 110 million and its rapidly developing industrial economy. For these reasons it developed a strategy of investing in mining projects abroad, to secure mineral supplies, establishing a strong smelting and refining industry at home and developing increasingly environmentally and economically efficient technologies to both commercialize abroad and to process metals profitably with least environmental degradation to its small and crowded islands at home.

Recent developments in the Japanese smelting industry can be divided into two phases:

(i) Late 1950s to late 1960s, in which old smelters were modernized and equipped with emission control technologies such as the dispersement of waste gas systems through high stacks, simple contact acid plants and electrostatic precipitors for dust removal. These controls were also installed in new smelters which were relocated to more isolated islands or peninsular tips.

(ii) Phase II began in the early 1970s when smelters were required to be adapted to meet new regulatory standards for emission control. These included a 'K value regulation' for SO_2 (a 'K' value is a specified value calculated from the emitted sulphur dioxide volume and stack height, and varies for different regions). In addition, standards were applied for the regulation of effluent water and tailing discharges, eg pH and toxic substance restrictions, etc. During this phase existing and new smelters installed double contact acid plants and/or SOx scrubbing systems, fugitive gas collecting systems, and waste water treatment systems.

Since Japan generally did not mine its own minerals either at home or abroad, it had less control over its quality and therefore its contained impurities, sulphur levels and ore grades. This set up an additional challenge to innovators in the smelting industry who had to respond to both rapidly evolving government legislation and changing characteristics of supply. The end result was increasingly to over-specify pollution control technologies – to go beyond existing legislation and develop technologies flexible enough to cope with variable qualities of mineral supplies.

European Community

The European Community is one of the greatest consumers of non-ferrous heavy metals in the world, and it accounts for a significant amount of processing of imported raw mineral materials. In total there are over 3,000 non-ferrous heavy metal works in the EC. Several hundred include roasting and sintering plants with a capacity of more than 100 t/y. Recent EC legislation now applies to all such plants which, in new cases, require licensing procedure prior to operation. Two types of emission values are considered relevant – values for specific heavy metals and values for dust emissions. Legally binding national emission limits are laid down only in Germany and Spain, and guidance values exist in the U.K. and Italy.

A key element of the relevant EC Directive is the concept of 'Best Available Technology Not Entailing Excessive Costs (BATNEEC)', to prevent or reduce air pollution. This equipment must be incorporated into a plant prior to the issue of an authorization to operate. Significantly, Article 7 of the Directive encourages individual members to exchange information about relevant measures and technologies to reduce emissions. This implies a special role for Eurometaux (Association of European Metal Producers). In this regard BATNEEC is interpreted as the technology which operating experience had adequately demonstrated to be the best technology commercially available to minimize emissions, providing it has been proven to be economically viable.

Applications for authorization will generally be deemed to meet BATNEEC criteria if the performance of the proposed plant, under normal operating conditions, is, in terms of atmospheric emissions, guaranteed to be equivalent or better than that of the designated BATNEEC. However, the Directive, after discussion, now gives individual Authorization Authorities the right to allow different interpretations in respect of the particular characteristics of the plant and its environment.

This gives industry the flexibility to incorporate any reasonable measures, provided the proposed plants meet the emission values and conform to broad principles of acceptable design practice. As a consequence of these rulings, the EC has now established a group of experts to designate BATNEEC pertaining to the production, storage and handling of heavy metals and to update the list as new techniques become available.

2.2 Regulation in Selected Developing Countries

Chile

Chile's experience has not been an absence of regulations, but rather a lack of concern and awareness of the environmental implications of its extensive mining activities. The regulatory agencies were apparently attached to the National Health Service, dissociated from the economic planners and staffed with untrained and unsupervised idealists.

Environmental standards were so stringent that they were unenforceable. For example, effluent water standards fixed maximum cation and anion concentrations higher than potable water standards; and workplace noise regulations were based on previous U.S. norms which have since been discarded as technically unattainable.

The enforcement of regulations has also been selective. For example, the only major private smelter in Chile – Chagres (operated by the Exxon subsidiary Cia Minera Disputada de Las Condes) – has been forced to meet standards more stringent than the current U.S. smelter emission standards, while only a few kilometres down the same valley the state-owned Ventanas smelter and refinery has operated for decades with neither sulphur dioxide nor arsenic removal facilities. Exxon has been forced to dismantle and transfer its 80-year-old tailings, at considerable cost, because they threaten to pollute Santiago's Mapocho River (this follows an earlier disaster when a hose with a thiophosphate collector ruptured at the San Francisco concentrator and contaminated half of Santiago's drinking water).

However, the major sources of mining pollution are state-owned (El Teniente, Andina and Ventanas), and the only action taken by the military authorities under Pinochet was to stop reporting the sulphur dioxide levels in the air and to reduce transport by 20%. The authorities concentrated on the visible carbon particles in the air, not the remedial measures of installing sulphuric acid recovery plants in Caletones and Ventanas which would have relieved the worsening infant respiratory problems in the city. The reduction of arsenic contamination from smelters remains Chile's greatest challenge. Fortunately, the new government is currently reviewing its environmental legislation.

Ghana

Gold is crucial to the Ghanaian economy, accounting for 20% of foreign exchange. The sector is growing thanks to a new Economic Recovery Programme, with World Bank facilitated credit, which is encouraging new foreign investment. In spite of the existence of an Environmental Protection Council (EPC), the mining industry has largely been governed by economic objectives. The EPC was underfunded, lacked skills and had little implementation powers. As a result, a growing extent of mining pollution has been reported – contaminated surface water from tailings (especially significant where piped water is rare), and forest defoliation due to the gases from roasting sulphide-rich gold ores.

Influenced by the World Bank's new green consciousness and growing awareness of potential health problems from pollution, in 1988 the EPC was asked by its government to prepare an Environmental Action Plan to make Ghana's economic development more environmentally sustainable. Six committees were formed to develop a National Environmental Policy, one of which was the Mining and Hazardous Chemicals Committee. The Committee was unique in the context of a developing country regarding the way it sought industry input and analysed the experience of other mineral-producing countries.

A seminar was held in late 1988 which identified a lack of environmental education within industry. This was followed by a study undertaken jointly by World Bank consultants and the University of Ghana which formed the basis, along with a study commissioned by the Ministry of Land and Natural Resources (also supported by the World Bank), for a regulatory framework and company guidelines which will minimize any adverse effects of mining on the environment. In the meantime, the government is requiring that any new project includes an environmental impact assessment.

China

Non-ferrous mining has traditionally been important in China, and its negative environmental effects are due, to a large part, to its inherent inefficiency and its use of obsolete technology and work practices. There exist reports of serious water and crop-land contamination from leaking tailings ponds, and of mercury pollution and land degradation due to small-scale alluvial gold mining operations.

Although China has an Environmental Protection Agency, and is considering a U.S. EPA model, it has issued no legislation for mining, nor are there any pertinent regulations in force or approved environmental practices on the part of the mining companies.

Bolivia

Until recently, environmental protection was subsidiary to industrial production. Increasing output to earn foreign exchange to finance government deficits, regardless of the costs involved in doing so, was the characteristic feature of the development of Bolivia's mining sector. This resulted in cumulative economic inefficiency, declining metal recovery rates and crippling company mismanagement. The inevitable consequence of this was growing environmental degradation of water, air and land resources with consequent health risks to local mining and peasant communities.

These factors, exacerbated by the 1985 tin price crash, led to the collapse of the state mining company, COMIBOL, which in turn had implications for environmental pollution since ongoing problems of acid mine drainage, tailings leakages and toxic dusts continued unabated despite mine closures or takeovers by private groups (with little access to credit or technology to improve the situation).

COMIBOL's poor record of environmental behaviour was in spite of the existence of 75 legal regulations, found in an array of codes, laws, rules and international agreements, which are under the jurisdiction of different government departments – Agriculture, Health, Industry – but not the Ministry of Mining and Metallurgy. There is no one institutional entity with responsibility for pollution control. Only recently have national concerns combined with pressure from agencies, such as the World Bank, started to change the situation. A recent effort by a number of agencies resulted in a Bill for environmental protection. However, last year the Senate objected to the Bill on the grounds that its implementation was impracticable.

Peru

Peru has been an important mining country since the turn of the century, with copper and silver and, to a lesser extent, zinc and lead amongst the economy's primary exports. The mining sector comprises three major groups – international firms such as Asarco (Southern Peru Copper Corporation), state mining companies (Centromin, Minero Peru, Hierro Peru) and a dynamic, medium to small-scale private national group. There also exist a large number of local equipment suppliers and consulting and engineering firms which could play an important role in promoting well-adapted environmental control technologies.

Environmental degradation in Peru has been relatively well documented, particularly following the formation of government agencies and non-governmental environmental pressure groups. Since 1984, the Office of Environmental Affairs has published 145 studies of environmental pollution 'events', many of which resulted from 'protests' made by affected peasant or mining communities.

However, despite the apparent existence of over 1,000 relevant pieces of legislation, most of the cases of environmental pollution continue unresolved. The major sources of pollution (tailings contamination of drainage waters and air quality pollution from smelter gases) are those receiving most explicit treatment, through the 1969 Water Law (DL 177752) and the 1962 Air Quality Law (DL 14084). More recently, following growing international as well as national concern, Peru has developed a Code for the Environment and Natural Resources (DL 613).

Although DL 613 does not reflect any substantive regulatory innovation, it has been important in putting all existing norms and laws within one overall legal document. Peru is probably unique amongst developing country mineral producers in that it has possessed both a long history of mining production, which has given rise to strong indigenous capabilities, and a relatively democratic political system, which may help to explain why there has emerged a tradition of environmental lobbying and protest.

Papua New Guinea

The history of environmental regulation in Papua New Guinea follows closely that of Australia. It was first defined in the country's constitution in 1975: 'We declare our fourth goal to be for Papua New Guinea's natural resources and environment to be conserved and used for the collective benefit of us all, and be replenished for the benefit of future generations'.

In 1974 an Office of Environment and Conservation was established within the Department of Natural Resources. However, placing this office in a department primarily concerned with resource development reflected a failure to recognize the distinction between regulation and investment promotion. A white paper was produced in 1976 which was the basis for three new acts in 1978: the Environmental Planning Act, the Environmental Contaminants Act, and the Conservation Areas Act. These were implemented from 1980 on the basis of guidelines issued by the Office of

Environment (OE). However, from the outset the OE was underfunded and understaffed, and came into conflict with more powerful ministries such as the Finance, Planning, and the Minerals and Energy Departments which previously were exclusively responsible for mining projects.

The history of large-scale mining in PNG began in the late 1970s with the development of the giant Panguna copper mine in Bougainville and the subsequent discovery of OK Tedi – a massive gold and copper deposit. The former was developed in an atmosphere of regulatory *laissez-faire* towards environmental issues, while, partly as a result of detrimental environmental effects from Panguna, the OK Tedi mine has been developed and operated within a far stricter agreement. This was imposed on the international consortium, led by Australian company Broken Hill Proprietary, by the PNG Government. The importance of the mine is reflected in the fact that by 1989 it contributed 42% of the nation's export earnings.

However, from the outset there were difficulties in regulatory implementation. For example, the consortium negotiated a budgetary ceiling (of $US180,000) on the environmental impact analysis which in reality meant that the survey only relied on secondary sources. Since many of the environmental obligations were open-ended, and could therefore form contingent liabilities, the regulatory framework affected the risk assessment of potential lenders. It also emerged that in some instances it may have been preferable to set arbitrary environmental standards rather than suffer delay and potential inconclusiveness trying to establish standards scientifically.

Since beginning its operations, OK Tedi Mining Ltd and the PNG Government reached an agreement for long-term joint environmental management based on standards determined by the Acceptable Particulate Level, which sets norms for suspended sediment downstream of OK Tedi. This has involved extensive scientific collaboration between the company's staff, the Australian Institute of Marine Sciences, the Institute of Soil Fertility and the Delft Hydraulic Laboratory in the Netherlands. The success of this will need to be reviewed in case study research. Finally, it is interesting to note that the company was able to increase its overall metal recovery through controlling the natural bacterial leaching of waste and reprocessing the lower grade ores.

CHAPTER THREE

Environmental Aspects of Non-Ferrous Mining

Heavily edited summary of a report, by the same name, published in 1992 by the UN/DTCD. A draft of this report was made available at the Berlin Round Table.

3.1 Potential Effects on the Environment

3.1.1 Water pollutants

Even though damage to human health due to mining-water pollutants is rare, in some places lethal pollution levels are experienced by aquatic organisms. The pollutants may occur more commonly at a level which causes some change in aquatic life but no total destruction.

Many metals, such as copper and zinc, are necessary to health in small concentrations but are highly toxic when present in excess. Depending upon their concentration, heavy metals can be lethal to fish, prevent reproduction or enter the human chain by accumulating in fish tissue, as for example cadmium and mercury (which may occur in zinc sulphide ores). The toxicity of heavy metals in fresh water is not only dependent on metal concentration, but also on other factors such as pH, water hardness and the occurrence of other metals. For example, toxic levels of copper and zinc for fish are dependent on water hardness:

Water hardness (mg/l CaCo)	Cu (mg/l)	Zn (mg/l)
10	0·005	0·3
50	0·02	0·7
100	0·04	1·0
500	0·11	2·0

Acidity which may result in an increased concentration of dissolved heavy metals can aggravate the problem. However, factors such as the buffering capacity of the receiving stream and the nature of the acid are also important in the overall impact of acid-containing effluents.

Thiosulphates may create environmental problems through their oxidation to acid in receiving waters, while the cyanides are lethal to many fish at very low concentrations (as little as 0·04 mg/l CN for trout). Some of the organic reagents used in base metals' ore processing operations may also be toxic. However, the concentration of most of the elements detected in effluents from concentration tailing ponds are usually well below the threshold of toxicity.

Oil forms a thin film over the water surface and can interfere with the re-oxygenation of water. It may also coat the gills of fish. Many water systems contain varying concentrations of suspended solids. However, depending upon their nature and concentration, suspended solids may interfere with self-purification of water by diminishing light penetration and hence photosynthesis reactions. In extreme cases, damage by silting includes water flora and fish destruction, flooding and interference with navigation.

An effluent containing a mixture of contaminants may have an overall toxicity which is different from that of its individual components. Biological testing of the effluent could be carried out to identify its potential for environmental damage.

3.1.2. Air contaminants

Dust elements in suspension in the air are potentially of concern to human health. Fall-out of particles causes contamination of soils, vegetation and water. The precise effects depend on the nature and the concentration of particles which are deposited. Gaseous pollutants related to base metals mining are of concern mainly to mine workers.

3.1.3 Other effects

The excavation of large quantities of material in open-pit operations, disposal of wastes above ground, erection of mining facilities, etc, may have a significant effect on aesthetic values. However, mining development, by its very nature, has the potential to unearth paleontogic fossils and archaeologic artefacts.

Population growth and infrastructure development are commonly associated with mine development. The positive or negative perception of these effects is largely determined by the degree and success of planning for that population growth. There may be significant positive impacts such as better health care, economic benefit to local population, etc.

Potential impacts to the environment have to be specifically studied when a mining project is to be developed in a remote and relatively unknown area; it should be insisted that environmental specialists and consultants be used throughout the planning and development stages of the project.

3.2 Potential Sources of Contaminants

3.2.1 Water pollution

The major and most significant source of liquid waste in the non-ferrous metal mining industry is acid mine drainage. Acid mine drainage is common in areas where mine openings intersect the water table and where the rocks contain iron sulphides (pyrite and/or pyrrhotite) or, less commonly, certain other sulphides. Where such pyritic ores are mined, rainfall leaching of rock waste stockpiles may be responsible for environmental damage. Less significant as a source of waste is the amount of suspended solids in mining operation drainage.

The volume and quality of mine water vary widely depending on the type of mining operation, geochemistry and hydrology of the specific mine. For open-pit operations, the volume of water depends on the ingress of groundwater and the precipitation into the surface area of the pit. The water is normally collected in sumps in the pit bottom.

For underground operations, the volume of mine water is influenced by the local hydrogeologic conditions, the infiltration of surface water into the mine, water used for drilling and dust suppression and any water contained in tailings backfill. The volume of mine water varies widely from mine to mine and even at a particular operation. Mine water is pumped to the surface from collection sumps where most of the suspended solids are decanted. Chemical characteristics of mine water may also vary widely depending on the mine.

Tailings from beneficiation plants are in most cases composed of a slurry which contains particles of ground material (gangue minerals and minor amounts of valuable minerals) in suspension.

The contaminants in tailings pond effluents include suspended solids composed of elements of the ore treated in the plant, heavy metals in solution, thiosalts and chemicals used for the treatment. The characteristics of effluents are highly dependent on the specific operation, and each mining project should be considered as a particular case. A brief review of tailings pond effluent contaminants is given hereafter:

Suspended solids: Provided the tailings disposal facilities are properly operated, the amount of suspended solids in the effluent remains at a very low level. Turbidity may arise from the non-settleable fraction of the suspended solids load. It may either be a true colloidal dispersion or very fine suspension of coarser particles; in any case, the effects of turbidity on fish life appear to be relatively small.

Heavy metals: A source of metals dissolved in tailings water may be mine water when it is used as process water, and/or flotation reagents such as zinc sulphate, copper sulphate used in lead-zinc beneficiation, and potassium dichromate in copper-lead-zinc ores treatment.

Thiosalts: These originate mainly in grinding and flotation of a number of sulphides under alkaline conditions. Thiosalts are of concern because they generate sulphuric acid as further oxidation forms the more stable sulphate ions. Presence of thiosalts in tailings pond effluents at concentrations of a few hundred milligrams per litre or less can create serious environmental problems through their oxidation to acid in receiving waters.

Process reagents: Reagents used most frequently in base metal flotation plants and in gold cyanidation plants are listed below. When the process is properly operated, most of the chemicals used for flotation, except pH modifiers, are absorbed at the surface of minerals. It may happen that some of them, when used in excess, remain in solution in the mill tailings. Most of the organic chemicals

Minero Peru's Cajamarquilla refinery. $ZnSO_4$ filtered solution is sent to 'champagne glass' for sedimentation before electrolysis.

in solution in the tailings slurry are oxidized in the tailings ponds and are not found in the final effluent. The kinetics of this oxidation are dependent on the temperature, and organic chemicals concentration in the effluent may be of concern in certain conditions during cold seasons.

Flotation reagents used in base metal concentrators

Acids:
Sulphuric acid

Alkalis:
Lime
Sodium carbonate
Sodium hydroxide

Modifiers:
Copper sulphate
Sodium cyanide
Zinc sulphate
Sodium sulphide
Sodium silicate
SO_2
Starch

Collectors:
Potassium amylxanthate
Potassium ethylxanthate
Potassium isopropylxanthate
Aniline Dicresyl Dithiophosphate
Gas-Oil
Amine

Frothers:
Dowfroth 250
Hexylic Alcohol
Pine Oil
HBTA frother

Cyanide is a reagent mainly used in lead-zinc ores flotation and for gold beneficiation. The amount of cyanide used in flotation operations is low enough to allow its removal from flotation tailings by natural degradation. The principal mechanism involved in natural cyanide degradation is the loss of volatile hydrogen cyanide.

In gold mills, although cyanide consumption is much higher, removal of cyanide by natural degradation is the method most commonly used. Prolonged exposure of wastewaters in the tailings pond is required. Where this is not economically feasible and/or where an acceptable effluent concentration of cyanide is not achieved, further treatment is necessary.

Run-off after rain can give rise to pollution problems. The disturbed land caused by open-pit mining is usually susceptible to erosion and silting may be a widepsread result. Roads used for ore and concentrates transportation, ore storage areas, non- or partially covered workshops, when subject to rainfall, may also become significant sources of contaminants in water courses if adequate control means are not undertaken.

It is recommended that surface run-off be diverted away from mining areas and surface facilities. If conditions permit, the different wastewater flows — mine water, surface run-off, tailings from the beneficiation plant — can be collected in a single point for a possible treatment before release into the environment.

3.2.2 Air pollution

Outside of the mine itself, and with the exception of relatively high production of dust in some specific areas, the non-ferrous metals mining and beneficiating industries have negligible air pollution problems compared with metallurgical industries.

Open-pit mining produces some carbon monoxide, oxides of nitrogen and sulphur (blasting and diesel-operated equipment). Only in very rare cases have these contaminants been known to cause external problems.

Dust particles are the main concern in external air pollution related to mines and concentrators. The main sources of dust in mining operations are listed below:

Point sources	**Dispersed sources**
Ore and waste loading points	Waste dumps
Ore chutes in haulage systems	Ore stockpiles
Screens in crushing plants	Haulroads
Exhaust from dedusting installations	Tailings disposals
Dryer chimneys	

3.2.3 Solid wastes and residues

Waste material located on top of the ore deposit which is to be exploited by surface mining methods must be removed first. The stripping ratio, i.e. the ratio of waste removed to ore tonnage, depends on the geology of the orebody and on economic factors such as recoverable value per tonne of ore and production costs. 2:1 is an average value for non-ferrous metals open-pit operations; however, much higher values, 5:1 for example, may be experienced in specific cases.

Overburden material removal is usually achieved by classical surface mining methods: shovels and trucks, draglines, bulldozer. Waste is stored in piles located close to the open-pit. Material characteristics are those of a run of mines ore, with blocks up to 1 m^3. Environmental concerns of overburden stockpiling include visual impact, pile stability, dust emission, and solid transportation and metal solubilization in run-off waters.

Underground mining methods produce material which may be used as backfill. Otherwise, waste has to be hoisted out of the mine and stockpiled by methods similar to open-pit waste storage.

Tailings of non-ferrous metals beneficiation plants are composed of a slurry which most frequently is discharged in a specifically designed containment area.

Tailings disposal methods:

1. Subaqueous discharge into tailings pond. Although lower *in situ* densities than with subaerial methods are produced, transfer of oxygen to tailings is inhibited, which is a great advantage when acid generating tailings are concerned.

2. Layered method. Tailings slurry is deposited in thin layers of uniform thickness (10-150 mm). The slope of the deposited layers depends on slurry characteristics and varies between 0·5 and 1·0%. Once a section of deposit has been covered with the desired thickness of fresh tailings, discharge is started on another section of deposit and the newly deposited layer is left to settle and dry from

several hours to a few days. The final configuration is a gently sloping mass of tailings, made from uniform layers with a resulting vertical permeability several orders of magnitude lower than its horizontal permeability. The advantage of the method is mainly that tailings are well drained and fully consolidated upon mine decommissioning. Conversely, it requires a high capital cost due to site preparation.

3. Thickened tailings disposal. Tailings slurry is thickened prior to its discharge from one or more spigotting points within the tailings disposal area. Tailings form a cone with a slope ranging between 2 and 8%. Typical percent solid required to achieve a slope of 6% range from 55 to 75%. This method enables a greater storage capacity within a given disposal area than with a conventional method as long as a high concentration may be obtained after thickening; it does not require the large dams associated with conventional disposal schemes, but the need to thicken the slurry results in higher operational costs than conventional methods.

4. Tailings disposal behind a dam. Conventional methods of tailings disposal consist of discharging the tailings behind a dam. The coarse fraction of tailings is usually used to build the dam. The tailings slurry can be discharged from a single point, through a series of spigots from a header, through spray bars, or through cyclones for the mechanical separation of tailings sands for the dam construction. Discharge of the tailings through cyclones should be preferred because a greater degree of control over the size distribution of sands used for dam building is provided.

A summary of tailings dams main construction principles is given hereafter.

(a) Upstream methods. This technique was used extensively in the past. It is now avoided because of the difficulty of ensuring its stability, particularly under seismic conditions. A typical upstream section incorporates the finely divided slimes fraction of the tailings into the structure, resulting in a heterogenous dam susceptible to failure.

(b) Downstream methods. This dam is built with the coarse fraction of tailings. The method provides the opportunity to design and construct a dam capable of meeting acceptable geotechnical specification, provided that a cyclone-classified sand fraction is used, the slope of the dam is adequate and the sand mass is properly drained. The major disadvantage of this method is the large volume of sand required.

(c) Centre line method. As in the previous method, the dam is built with the coarse fraction of tailings. The dam centreline is maintained in the same vertical plane as the height increases. Downstream of the centreline, the dam has the same characteristics as in the downstream methods; therefore an acceptable structure may also be provided.

Whatever the dam construction method used, the most common failure associated with tailings impoundment structures are those related to spillways, decants and diversions. If the hydraulic capacity of these facilities is exceeded during periods of peak flow, a failure in the dam structure can result, leading to the liquefaction and release of the stored tailings.

All tailings impoundments must be evaluated at the design stage to ensure that rainfall/run-off can be accommodated through storage, diversion around the facility or discharge via a spillway or decant structure. In establishing a design criteria, a statistical assessment of the occurrence of an inherent run-off event should be made. Events which occur only once in a thousand years have been used in some cases.

Where potential failure of a tailings system has life-threatening implications, upper limits should be used, although it is highly recommended not to install any tailings disposal facilities where such events could occur. Devices used to release excess supernatant water from tailings disposals include decant towers, siphon systems and barge-mounted pumps. In the latter case, tailings systems must be designed so that peak run-off events can be safely stored within the impoundment and subsequently released.

3.2.4 Subsidence

The removal of material by underground mining creates a potential for ground movement and consequential deformation of the surface. The circumstances under which this may arise vary widely. The main parameters are:

- the geometry of the mineral deposit;
- the method of mining;
- the nature of the mineral deposit and the overlying strata.

In many cases, application of rock mechanics enables a fairly reliable qualitative prediction of subsidence. Damage occasioned by subsidence may include major fracture of the ground surface which can cause severe damage to buildings, and/or continuous surface deformation, usually in the form of a surface trough. Uniform displacement seldom causes major damage; problems are more often occasioned by differential displacement (for example, interference with groundwater flow and changes in the gradients of road, gas pipes, etc).

3.2.5 Noise and vibration

The major sources of noise in the mining industry are fixed plant, mobile equipment and transport movements. Examples of noise levels associated with various fixed plant installations are given below:

Equipment	Noise level (dB(A))	Measurement location
Electrical ventilation fans	90-100	5 m
Compressed air fans	up to 110	5 m
Jaw crusher	90-100	Operator position
Cone crusher	92-98	Operator position
Compressed air hammer	104-112	Operator position
Drill sharpeners	102-122	Operator position
Ball mill	up to 100	Operator position
Flotation equipment	63-91	Inside flotation building

Mobile equipment used on site is associated with drilling, blasting loading and haulage operations. Some examples of noise levels are given below:

Equipment	Noise level (dB(A))	Measurement location
Compressed air rock drill	110-115	1 m
	98	15 m
Large portable compressor	80	7 m
7 m^3 dragline	90-92	Operator position
Diesel trucks	74-109	Driver position
	88	15 m
Electric shovels	78-101	Operator position
Graders	76-104	Operator position
Dozers	84-107	Operator position
	87	15 m
Locomotives	75-95	Driver position
Rotary drills	72-100	Operator position
Front end loaders	83-101	Operator position
Scrapers	92-104	Operator position
	88	15 m

The noise levels given in these tables are measured close to equipment. When dealing with external nuisance, a reduction factor has to be taken into account. From different mines' experience, a continuous noise level exceeding 75 dB (A) is unusual at the perimeter of the mining property. It is usually accepted that a 50-70 dB (A) is a maximum acceptable level.

Air blast is the term used to describe the air vibrations generated by blasting operations. The main factors influencing air blast effects include the type and quantity of explosive, the degree of confinement, the method of initiation, climatic conditions, and local geology and topography.

Air blast waves may give rise to damage and nuisance. The table below indicates effects of air blast overpressure upon structures.

Overpressure (lb/in^2)	(g/cm^2)	Structural effect
0·03-0·05	(2-4)	Loose window sash rattles
over 0·1	(7)	Failure of stressed or badly installed window panes
over 0·75	(52)	Failure of correctly installed window panes
2·0	(140)	All window panes fail
over 2·0	(140)	Plaster cracks

However, normal mine blasting results in overpressures well below those indicated above. The damage potential of air blast is largely confined to structurally unsound buildings, bad blasting practice and unusual atmospheric effects.

In blasting operations, an undesirable result of detonation is that the surface of the ground in the vicinity of the blast undergoes displacement, whose amplitude depends on the distance from the blast, the energy released in the explosives and the local geological conditions.

Among the factors used to assess the effects of ground vibration, the peak particle velocity appears to be the most closely related effect upon damages. The frequency of the vibration has also an effect upon damages resulting from blasting: the lower the frequency, the greater the damage for a given peak particle velocity. The table below gives indications of the effect of ground vibrations upon structures.

Peak particle velocity (in/s)	(mm/s)	Damage
below 2·8	(70)	Nil
4·3	(110)	Fine cracking and fall of plaster
6·3	(160)	Cracking
9·1	(230)	Serious cracking

3.3 Acid Mine Drainage

Acid mine drainage (AMD) may be defined as the inorganic chemical water pollution resulting from the oxidation of sulphide-containing minerals, mainly pyrite (FeS_2) and pyrrhotite (FeS). Consequently, acid effluents containing elevated levels of dissolved metals (including iron, zinc, cadmium, manganese and lead) are produced.

The main deposits concerned are base metals sulphide ores. Coal deposits which contain a significant amount of pyrite may also produce AMD. AMD is a problem which takes place while the mine is in operation and also after closure. However, control measures are usually not the same for both cases. In a closed mine, AMD can affect not only underground waters but also waste rock stockpiles and treatment plant tailings.

Three stages are considered in the generation of AMD:

Stage 1. This involves the relatively slow chemical or biological oxidation of pyrite and other sulphide minerals near neutral pH, producing ferrous iron and acidity. This step may be catalysed by the bacteria Thiobacillus ferroxidans through direct contact with sulphide minerals.

Stage 2. Ferrous iron is oxidized into ferric iron which precipitates as ferric hydroxide and releases more acidity. As the pH falls even further (below about 3·5) ferric iron remains in the solution and oxidizes the pyrite directly.

Stage 3. Bacteria rapidly catalyzes the process by oxidizing ferrous iron into ferric iron and the overall rate of acidity production is increased by several orders of magnitude. Large quantities of acid are produced. It may be seen that, while the abundance of sulphide minerals is an important factor, both oxygen and water are required reactants, bacteria acting as catalysers.

Water also plays an important role in removing the oxidation products from the surface of sulphide minerals: acid and metal sulphate salts may accumulate within a waste deposit during relatively dry periods and be released at times of higher precipitation. Carbonate minerals such as calcite and dolomite are the principal natural acid neutralizing components of mine waste. Those minerals can neutralize the acid produced by sulphate oxidation and prevent the establishment of low-pH environments required by micro-organisms. Under these conditions, AMD cannot develop.

3.4 Closure of Mines and Abandoned Sites

When a mining operation shuts down, measures should be taken in order to reclaim the site and, at least, to ensure safety of persons and domestic animals. Quite often, old mines were abandoned without adequate reclamation measures having been undertaken in due time.

One of the main parameters that determines the measures to be undertaken is the possible acid-producing character of a mine site. At non-acid-producing mines, reclamation measures may be instituted such that no long-term waste water treatment facility is needed. Conversely, at acid-producing sites, contaminated drainage can occur, resulting in a long-term treatment obligation if it is not effectively controlled.

Drainage water may come from mine, waste stockpiles or tailings disposal facilities. At non-acid mines, mine water usually meets environment requirements. Stockpiles and tailings impoundments have to be treated so that surface erosion by water is prevented.

At acid-producing sites, the final objective of post-operational measures is to reduce the volume and strength of acid-drainage to a level where passive techniques are a sufficient treatment of the residual drainage before discharge. However, such an objective is not usually achieved and a long-term treatment of acid drainage water has to be maintained, at a cost which can be very high.

For obvious safety reasons, all accesses to underground workings must be closed carefully. When treating an abandoned mining site, location of accesses to ancient workings may be a problem. Shafts were usually sealed with wooden plugs topped with some fill or more recently with concrete caps. Nowadays, it is recommended to completely fill the shaft with inert material. The entrance to adits should be plugged with concrete. Long-term subsidence shall also be evaluated to determine if surface deformation could result in damage to buildings.

Base metals open-pit operations are frequently very large and backfilling with the waste overburden previously removed during mine operation is not economically feasible. Rehabilitation of such sites is then difficult, although, in some cases, abandoned pits are used for water storage purposes or recreation.

When non-acid generation conditions prevail, the main issues during the post-operational phase are the geotechnical stability of the pile (and/or the impoundment) and stabilization, usually by vegetation, to prevent wind and water erosion.

3.5 Control Technology

Methods that can be used to control the different sources of contamination in the non-ferrous metals mining industry are numerous. It should be borne in mind that no general rule exists, and that the constraints imposed by each specific site will dictate the option that can be chosen. Flexibility in the approach to the design and performance of project facilities is critical to achieving controls that are technically and economically feasible.

3.5.1 Water pollution

Successful control of water pollution is based on the knowledge of the quantities and qualities of all waters which may be affected by mining activities, the quantities of water required during mining and beneficiation, and the qualities of the process waters after use.

Unless a properly designed monitoring program has been carried out, any control measures which are taken are unlikely to result in the desired result at the optimum cost. The main features that should be contained in a monitoring program include a definition of objectives, selection of parameters to be measured, sampling locations and procedures.

Main components of a monitoring program:

Physical

- Temperature
- Turbidity TDS
- Water flows

Chemical

- Conductivity
- Alkalinity
- pH
- Hardness
- Colour
- COD/BOD
- Nitrogen
- Phosphorus
- Metals

Biological

- Phytoplankton
- Zooplankton
- Benthic organisms
- Fish
- Water fowl

The physical control of volumes and routes of water at a mine site is a major task. The volume of water used in mining and beneficiation facilities should be minimized to prevent contamination of unpolluted water, and any polluted water flow should be intercepted and diverted to an appropriate place for possible treatment.

Where possible, it is recommended to collect all mine exhaust water in a single point from where it can be directed to a treatment facility or to the environment (if its chemical characteristics allow). However, the opportunity to reduce the volume of underground mine water is limited: the main methods are to seal and grout old bore holes and to reduce water penetration through shaft linings.

Water control in surface mining operations essentially depends on the site topography. Where the open-pit is developed under the topographic surface, run-off water should be collected at the lowest point. Where surface mining operations are conducted on hill sides, the purpose of water control is to limit the amount of run-off water coming from higher up the slope. Interception ditches are frequently used. Run-off waters flowing over the mining site are collected downstream in ditches and directed to decantation ponds installed behind draining dams.

Liquid effluent controls in beneficiation plans mainly consists in collecting any liquid spillage within the plant in one or several sumps and directing this effluent with the tailings towards the tailings disposal system.

Reagents used for the process, including flotation reagents and cyanide in gold treatment plants, require special care within the plant and in storage facilities. Holding tanks and distribution pipes have to be properly designed to prevent any spillage. Reagent stores and preparation plants should be equipped with a concrete sump of sufficient capacity to contain the total quantities of stored reagents, at least for highly toxic product such as cyanide, to avoid any release in the environment in case of leakage.

Though contaminated waters resulting from acid mine drainage can be readily treated, steps should be taken to reduce the volume and improve the quality of contaminated waters. Passive treatment methods should be preferred whenever technically feasible for obvious cost reasons. All techniques attempting to control one or more of the basic elements of acid generation — oxygen, water, bacteria and sulphide minerals — should be considered.

Prevention and control of acid mine drainage in a mine while it is in operation is often difficult because of the constant evolution of the workings. However, acid underground mine and/or open-pit waters must be neutralized and have their metals removed. This can be achieved in a treatment plant specific to mine drainage, or more commonly, in the tailings ponds. Removal of metals from dilute solution in a treatment plant results in the production of concentrated heavy metal sludges which should be carefully managed to prevent their introduction into the environment.

Precautions which can be taken as far as solid disposal is concerned include minimizing the exposure of sulphide minerals, avoiding the use of sulphide bearing materials for the construction of dams, etc, and designing and constructing adequate drainage control facilities.

An underwater tailings disposal prevents the oxidation of sulphides contained in concentrator tailings. However, few data concerning the long-term impact of such a practice on the quality of the water body used for disposal are, at present, available. In any case, a site-specific evaluation is recommended if such a disposal is taken into consideration.

When a mine has ceased operation, complementary measures can be taken. Minimizing the access of air to sulphides in underground closed mines, which is the most effective measure to reduce the rate of acid generation, can be achieved by flooding and/or sealing the mine.

Post-operational acid drainage control techniques that can be applied to waste and tailings disposals include; capping thc waste (with clay, plastics, grout, asphalt, etc), application of top soil and revegetation, and a combination of both preceding methods. The main disadvantage of these techniques is a cost which may be prohibitive when large waste dumps are to be treated.

Other control techniques are under investigation but have not proved to be successful at present: surface covering of lime or limestone placed on waste, use of bactericides to kill bacteria responsible for acid generation, and *in-situ* neutralization by addition of carbonate with waste. Recent studies have also demonstrated the value of natural or artificial wetlands to remove heavy metals.

The most common method of acid mine drainage water chemical treatment consists of neutralizing free acid and precipitating metals as hydroxides under alkaline conditions. As ferric hydroxide is far less soluble than ferrous hydroxide, waste water is aerated to oxidize ferrous iron which normally occurs in acid mine waters. For cost reasons, other chemical oxidants (for example ozone, hydrogen peroxide, chlorine, permanganates) are rarely used in the mining industry.

A consequence of the use of lime is a possible precipitation of gypsum (hydrated calcium sulphate). Precipitation of gypsum increases the quantity of sludge generated by waste water treatment and forms a concrete-like crust which may clog the equipment if not removed periodically.

The separation of precipitated metal hydroxides from the liquid phase is possibly the most difficult aspect of the treatment of acid waste waters. Metal hydroxides precipitate as amorphous solids with low settling rates and poor sludge dewaterability. Various methods are used to improve the dewatering of hydroxide precipitates. These include the recycling of sludge to the neutralization system and the use of polymeric flocculants. Flocculants are costly reagents and the exact amount to be used is to be determined as precisely as possible. This requires brench- and pilot-case studies.

Treatment sludges contain metals removed from waste waters. They are usually disposed of with mill tailings in separate disposal lagoons by burial in on-site landfills and in some cases in underground and open-pit workings. Whether the sludges are mechanically dewatered prior to disposal depends on the particular situation; in any case such a dewatering process is expensive. All such sludge or residue disposal sites are a potential source of pollution due to hydrological processes. Hence their design operation and eventual abandonment require careful consideration.

Waste waters from gold mining operations can contain cyanide in the range of 100 to 400 mg/l and should be treated. Biochemical oxidation and neutralization prior to discharge into the environment is the most promising process for removal of thiosalts, provided long residence time of waste waters in ponds can be achieved. Waste waters from gold-arsenopyrite mining operations may also require the removal of arsenic which can be very effectively removed by precipitation as either calcium or ferric arseniates; arsenide containing waters should be oxidized prior to arsenic precipitation. Coprecipitation of arsenic with ferric hydroxide at alkaline conditions is also used.

3.5.2 Air pollution

In the base metals mining industry, external air pollution is mainly due to dust. The amount of dust at a specific point can be measured with different equipment: standard dust deposit gauge, which provides an integrated record of total fall out over the period of exposure; directional deposit gauge to measure horizontally borne dust; or a smoke filter.

The most effective way to control dust emission from point sources is to prevent its formation by controlling the handled product's moisture content. Run of mine ore moisture depends on the hydrogeology of the deposit and mine water control. Concentrates produced in the mill are frequently filtered and filtration control provides means to control concentrate moisture and therefore its possibility to emit dust. Basically, an 8-10% moisture content prevents dust emission from sulphide ore concentrates.

Water, normally incorporating a wetting agent, is used to increase the surface moisture in places where dust may be produced; crushers, screens, ore chutes, etc. Water sprays are far more effective in preventing dust production than in knocking down dust in the air. An effective dust suppression system requires careful selection and positioning of the spraying equipment to enable pre-wetting with water.

An effective alternative at transfer points is to limit the height of fall of the product or to enclose it during its fall. Use of totally enclosed storage areas is frequently necessary to prevent dust emission by wind from finely ground products such as mill concentrates. In some cases, wetting the ore is impractical; removal of dust is achieved with a forced extraction system which draws dust-laden air from its point of origin into a separator. Most commonly used separators include mechanical collectors (eg gravity settling chamber and cyclones), bag filters, wet scrubbers, and electrostatic precipitation (being expensive, the latter is rarely used in mining industry).

The control of dust from dispersed sources is more problematic. Surface mining excavations and waste rock piles usually produce dust as long as they are in active use. Dust emission normally lessens rapidly after abandonment because the supply of fine particles is exhausted. Water sprays, which can be automated, have been found to be adequate.

While in operation the dust emission problems from tailings dams are similar to the previous ones. Chemical fixatives can be used for temporary dust control but, when abandoned, dust emission is more important because of the fine material from which tailings dams are made. As spraying an abandoned tailings dam is not economic, revegetation is usually the most effective way to control dust emission.

Haulroads, where dust emission takes place by spillage from trucks and by wheel abrasion, are a major but intermittent dust source. The most usual approach is to spray water on the roads. Control is facilitated on properly compacted and graded haulroads. Paved and coated haulroads enable a very efficient dust control; however, such measures are rarely practical for cost reasons.

Flotation concentrates are fine materials, usually with a high value. If no measures are taken, losses due to spillage can be important and costly, whatever the environmental concern. Open trucks or railroad cars should be avoided. Wholly-sealed containers are more commonly used.

3.5.3 Solid waste disposal

In selecting a site for tailings disposal the following characteristics have to be taken into account to ensure a durable stability of the tailings deposit; adequate capacity to store the quantities of tailings projected for the expected life of the mine (with some potential for expansion), and suitable geotechnical conditions of the underlying rocks.

Where borrowed material is used to build the dam, the rules applying to earth dam construction are to be used. Important features, as far as tailings dam stability is concerned, include homogeneity of the dam (built with the coarser fraction of tailings), close control of the separation between the coarse and fine fractions, installation of properly sized drains, permanent monitoring of the phreatic surface within the dam, control of possible seepage on the downstream slope of the dam, and close control of the dam slope (the maximum slope depends on the material used).

Numerous geotechnical considerations apply to waste rock disposals, including suitable geotechnical conditions of the underlying rocks, diversion of structural water around the storage area, proper drainage of the waste rock pile, and design of the pile slope as a function of the geotechnical properties of stored materials. In some cases, such as limonites in lateritic nickel ores, compaction is required.

The utilization of waste material reduces the environmental problems of conventional disposal. However, the large amount of material involved, the frequently remote location of mines, and the occurrence of undesirable elements (acid-generating minerals for example) are important drawbacks to a significant re-use of mining wastes and tailings.

The most widespread usc for tailings is as hydraulic backfilling for support purposes in underground mines. Usually, only the coarsest fraction is used, while the fine material has to be disposed of in a surface impoundment.

3.5.4 Subsidence

The alleviation of subsidence damage may be undertaken either by precautionary measures on surface to protect installations, or by modification of the mining method so as to minimize deformation of the surface. Precautionary measures on the surface include avoiding locating new installations in areas of geological discontinuity, designing new buildings whose structures are either completely rigid or completely flexible, and choosing flexible pipes and joints for pipelines.

Underground precautions which have to be taken vary widely, and depend on many parameters such as the geometry and the nature of the mineral deposit. They include leaving safety pillars and/or backfilling the excavated areas.

3.5.5 Noise and vibrations

Basic techniques are available to control noise in mining industry. These are aimed at reducing the noise energy generated at source, isolating the source, and by increasing the noise absorption between the source and the listener.

3.6 The Working Environment

In a world of rapidly changing technologies and products, including new chemical substances and growing concerns about hazardous wastes, issues related to the working environment have become increasingly complex and more closely linked to the general environment. Consequently, the past distinction between the working environment and the general environment has become less clear. Lasting solutions to the problems of the general environment cannot be tackled unless the problems of the working environment are satisfactorily solved. Public support for stricter environmental regulation has been getting stronger year by year and the environmental requirements being introduced by governments are focusing on making the working environment safer and healthier, and on improving the general environment by controlling the land that is available for exploitation, minimizing the disturbance of the land surface, maintaining air quality and protecting water resources.

3.6.1 Main occupational hazards and health effects

Safety and health issues in mining operations are in a number of ways similar to those in any industrial complex. However, due to the nature of a mining operation, there are a number of areas which require special consideration. For example, by its very nature, mining involves the handling, transportation and processing of large quantities of material. Thus, the chemical composition of the material, its concentration in the working environment and control of some parameters are critical to assure workers' health.

Airborne dust is a hazard to the working environment in

underground and surface mining operations. It can also be of concern in beneficiation plants, especially in crushing operations. Among the most hazardous of the major metals, in terms of occupational risk, mainly by inhalation, are lead and nickel. Other secondary contaminating metals must also be addressed: cadmium (associated with zinc and lead ores), arsenic (mainly associated with gold ores) and mercury (associated with many types of polymetallic ores). To these metals must be added the possible presence of crystalline silica-quartz (usually found in dust resulting from hard rock mining).

The risks associated with these materials are fundamentally determined by the chemical composition and the physical and physicochemical properties of the dusts particles (including the mineralogical composition). As far as metals are concerned, great care must be taken in identifying the 'substances' present. That is generally possible through mineralogical analysis but this analysis must be made on the actual dust and not only on the original rocks. In other words, 'speciation', i.e. the relationship between the health effects of a metal and the chemical composition of the compound in which the metal occurs, must be taken into consideration.

'Speciation' requires a consideration, in terms of mechanisms of possible biological reactions, of the concept of solubility (meaning not only solubility in water but also in biological fluids like blood). For insoluble particles, which act on the innermost parts of the lung, only those particles small enough to penetrate and deposit in the pulmonary spaces, such as silica dust, are of interest for the assessment of the health risk. The best understood health effects associated with exposure to metals compounds in the mining industry are of the systemic intoxication type and are reported for lead (from oxidized ores) and mercury.

Chronic exposure to free crystalline silica dust may lead to silicosis, a pneumoconiosis which is a very serious, progressive and irreversible disease. Since the composition of ores and surrounding rocks varies, other hazardous agents can occur, therefore it is important that, for each case, the constituents be determined and their toxicology investigated.

Gases which are often present in underground mines and which have to be carefully monitored are carbon monoxide and nitrogen oxides. Diesel engine exhaust gases and nitrogen oxides contain significant amounts of heavy hydrocarbons, aldehydes, ketones, sulphur dioxide and carbon dioxide. When urethane foam systems are used, there can be a risk of exposure to diphenyl methane diisocyanate. Nitrogen oxides and ozone can also accumulate when electric welding is performed in confined spaces.

Methane is not common in base metal mines, but carbon dioxide resulting from decay of timbers in abandoned mines or emanating from rock strata is not uncommon. Decaying organic material can also lead to hydrogen sulphide and to oxygen deficiency. Aldehydes and sulphur dioxide are irritants to the upper respiratory tract; heavy hydrocarbons, for example benzo-pyrene (fine aerosol), can even be carcinogenic.

Some chemical reagents used for the beneficiation of base metals by flotation are very toxic. Cyanide is used for the froth flotation of base metals and for gold processing. Mercury is still in use in some gold processing plants, particularly in small-scale artisanal gold mines. Both are extremely toxic. Other harmful agents include organic collectors, such as xanthates. A very important characteristic of cyanide chemicals is that they can penetrate through intact skin, therefore skin contact as well as inhalation must be avoided.

Miners are often exposed to high noise levels. Continuous exposure to intense noise, at levels exceeding 85 dB(A), may cause hearing loss. With prolonged and repeated exposure to high noise levels, hearing loss may become permanent and irreversible. Evidence, from workers with hearing losses resulting from occupational exposure to noise, shows that such losses start and are most prominent in the frequencies around 4000 Hz. As the hearing loss progresses, it involves other frequencies to the point that the person becomes permanently deaf.

Noise levels, as specified by EEC regulation, are for a daily exposure level of 85 dB(A) with no peak sound pressure exceeding 140 dB(A). Regulations in different countries vary but are usually between 85 dB(A) and 90 dB(A) for an 8-hour work day, with higher levels allowed for shorter periods.

Health problems due to heat stress are very common in the mining industry. The main sources of heat in mines are from the adjacent rock walls (the increase in rock temperature with depth can reach as much as 1°C per 150 m), adiabatic auto-compression of the intake air, machinery and human occupancy.

The four environmental factors which determine heat stress are: the temperature, humidity, velocity of the air and radiant heat. In many operations in the mining industry, combinations of these factors may result in serious heat stress to the workers, who may be performing heavy work and producing large amounts of body heat, thus adding appreciably to the heat stress problems. Air temperature and relative humidity affect thermal comfort in inverse proportion. For instance, 100% humidity at a temperature of 30°C are practically insupportable conditions.

3.6.2 Environmental monitoring

In order to establish the extent of exposure to occupational hazards in underground and surface working areas, it is recommended to establish routine monitoring programmes.

The presence of air contaminants in the workplace can be assessed through techniques of air sampling and analysis. The concentration of airborne contaminants in the workplace air varies with respect to time and location, therefore an important aspect of hazard evaluation is to ensure that the sampling procedure is representative. For this, an adequate sampling strategy must be designed and followed.

Airborne particles can be evaluated by means of continuous sampling devices or by direct-reading instruments. (The latter are not recommended in the case of dust causing chronic effects since the results are instantaneous.) For airborne dusts, which may cause pneumoconioses, it is important to evaluate the 'respirable' fraction, i.e. the portion of the airborne dust which can penetrate to the pulmonary spaces.

There are different criteria for size separation of the 'respirable' fraction of dust. The most widely accepted are those recommended by the Medical Research Council of Great Britain and the U.S. Atomic Energy Commission, the latter slightly modified and adopted by the American Conference of Government Industrial Hygienists in 1988.

The frequency of sampling depends on the situation. An initial survey should be carried out and, if the air concentrations are found to exceed the adopted standards, control measures should be implemented. Evaluation should be made shortly after the implementation of control measures in order to evaluate their efficiency and thereafter routinely.

It is also necessary to analyse the airborne dust for chemical and mineralogical composition. In order to assess the risk of silicosis, it is necessary to evaluate the content of free crystalline silica. Dust containing lead, nickel, cadmium and arsenic compounds should be evaluated in a similar way. However, when dealing with a soluble compound, the total dust should be measured.

Unless a direct reading device is used for airborne contaminants in the gaseous state, the collected sample has to undergo further analysis. There are analytical techniques particularly useful for occupational hygiene applications. Direct-reading devices, with built-in alarm systems, are very useful for atmospheres which can be immediately dangerous to life (for example, where there are excessive concentrations of carbon monoxide). In fact, there are instruments which combine alarms both for oxygen deficiency and for carbon monoxide.

CHAPTER FOUR

Small-scale Mining in Developing Countries

Extracts of a report by the U.N. Economic and Social Council's Committee on Natural Resources. This work was tabled for discussion at the Berlin Round Table.

In September 1988 in Ankara, the Inter-regional Seminar on Small-scale Mining in Developing Countries was convened by the UN/DTCD in conjunction with local executing agencies. Papers presented and statements delivered by more than 50 participants from developing countries clearly confirmed the high level of small-scale mining activities in a number of countries, and the firm interest to promote growth of the subsector. Another recent event of significance was the establishment in 1989 of an international agency for small-scale mining, Small Mining International (SMI), at the Ecole polytechnique in Montreal, Canada. SMI is a non-profit organization specifically dedicated to strengthening and supporting the small mining sector as an aid to rural social and economic development, particularly in developing countries. Also in 1989, in India, the National Institute of Small Mines was created to promote the role of small-scale mining in national development and international co-operation. The principal objectives of the organization are to encourage the small mining sector to use technical and scientific methods in its operations, and to raise consciousness regarding environmental protection.

4.1 National Initiatives

4.1.1 Zimbabwe

Zimbabwe has an impressive small-scale mining record, and a Small-scale Miners' Association has existed since 1982. As a non-governmental organization of operating and prospective small miners, its main goals are to function as an information and advisory centre and to provide technical and business management training. The country has a long tradition of small-scale mining. Chromite mining in the Great Dyke district dates back to the turn of the century. The local chromite mining industry today is controlled by a subsidiary of Anglo American Corporation. In 1983-84, when the chromite industry went through a particularly difficult period, the less profitable smaller operations were abandoned.

Faced with unemployment, mine workers, with the support of the government, began to establish chromite mining co-operatives and continued mining operations. Since then 28 chromite co-operatives have been established. The formation of the mining co-operatives was facilitated by the Zimbabwe Mining Development Corporation Act of 1982. The Act states that it shall be one of the duties of the Zimbabwe Mining Development Corporation (ZMDC) to encourage and undertake the formation of mining co-operatives.

The co-operatives receive extensive technical and administrative assistance from the government through ZMDC. Services provided primarily include assistance in negotiating ore prices, providing geological support in prospecting, attaching resident advisers to each group of four co-operatives, giving administrative and technical training to the management committees and assisting in the acquisition of equipment. A small fee amounting to about 3% of sales revenue is charged by ZMDC for these services.

4.1.2 Bolivia

Between 1985 and 1986, in an effort to streamline its operations, the State-owned mining company COMIBOL laid off more than

28,000 workers. Most of them subsequently joined mining co-operatives, whose total workforce had reached 40,000 by the end of 1989. Co-operatives are legally working on former COMIBOL tin and base-metal mines abandoned because they were considered economically unattractive. In addition to tin, six minerals are also recovered: gold, silver, zinc, antimony, lead and tungsten. In 1989, the co-operative production contributed 37% of the total foreign earnings and almost 19% of Bolivia's total exports. They produced 9·6 t of tin, more than twice as much as COMIBOL. In spite of the fact that the mining artisans use rudimentary tools and methods, and get little or no technical support from COMIBOL, their production was 57% greater than in 1988. By contrast, the Mining Bank, Banco Minero, provides credit lines to the microscale gold miners.

4.1.3 Ghana

This West African country has been mining alluvial gold for many centuries and diamonds for several decades. Traditionally, a large rural labour force has been engaged in working the deposits individually on a very small scale, using artisanal methods. Few of the artisanal operations have been officially licensed.

An over-valued currency and restrictive legislation have in the past contributed to driving most of these activities into illegality. As a result, the majority of the precious minerals produced by the artisanal segment left the country illegally. According to official estimates, between $15 million and $30 million worth of gold and diamonds were smuggled out of Ghana in 1985 alone. Other sources estimate that the values were considerably higher.

Against this background, one of the objectives of the Minerals Commission, established by the government in 1984, was to create an appropriate institutional framework for the artisanal/small-scale mining segment. A process of legalizing the artisanal mining activities was initiated in 1989, with support from the World Bank. In the initial stage, district centres were established by the authorities in eight artisanal gold mining areas, responsible for registration, licensing, training and monitoring. Mine wardens appointed by the Mines Department advise miners in the use of more efficient mining methods. In addition, regular training programmes are offered to the artisanal miners.

As part of a purchasing system designed to eliminate illegal trading, buying stations are being installed in all districts. Field purchases of gold will be conducted by buying agents licensed by the Precious Minerals Marketing Corporation. The pricing arrangement envisaged allows payment of at least 90% of the world market price, based on the market exchange rate instead of the official rate. Buying agents also provide miners with essential operating supplies, such as sluice boxes, pumps and mercury. The regularization campaign of the government has been a complete success. The gold mining activities of more than 11,000 artisinal workers were legalized in less than one year.

4.2 Importance of Small-scale Mining

Conducted as a systematic and disciplined economic activity, small-scale mining has the potential to make a significant contribution to national objectives in developing countries in various ways. As opposed to large industrial operations, small firms have consistently been identified as a vehicle for the creation of an indigenous entrepreneurship which is a vital force in the process of industrial development. Lower barriers to entry in terms of time, capital and formal training requirements are the principal reasons for this favourable characteristic of small enterprises.

While modern mining is a highly capital-intensive business, the financial requirements of small operations are usually only a fraction of those of larger units. This is not only due to the lower capacity levels but also to the fact that small mines, being inherently labour-intensive, generally use mining techniques with a lower degree of mechanization. Thus, in small-scale mining, investment costs per job created are typically only 10-20% of those costs in large, fully mechanized operations, an attribute of particular relevance in economies in which capital is a scarce factor of production, as in many developing countries.

Similarly, small mines can be brought into production in a fraction of the time required for large operations. While the latter may need a time span of 5-10 years from the identification of the deposit to full production, small units will usually be operational in a period of from one to two years. Short development periods result in reduced amortization time, rapid return of the capital invested and the early accumulation of new funds for mine expansion or the implementation of additional projects.

Small-scale mining and prospecting conducted by small miners or local individuals have traditionally been an important factor in mineral industry development. It has been estimated that in the U.S., one of the leading mining countries in the world, more than half of all major mining properties active at the beginning of the past decade derived from exploration targets identified by small prospectors. In addition, small miners frequently play an important role as scavengers, extracting mineralizations too small in tonnage to be mined on a larger scale. Indications are that, world-wide, the number of such mineralizations exceeds the number of deposits mineable on a large scale by several magnitudes.

Small-scale mining can, moreover, induce favourable indirect effects in the form of linkages with the secondary and tertiary sectors of the economy. This has been confirmed by the multiplier effects determined in a number of states with significant small mining activities in the U.S. Typical values obtained for income multipliers related to small-scale mining and mineral processing amounted to 2-3·5; those for employment multipliers were in the range of 3-5, while output multipliers were found to lie between 1·3 and 2.

In view of its typical labour intensity, small-scale mining can have an important impact on the employment situation, particularly in economies with widespread unemployment or under-employment. Labour intensity is greatest in the artisanal operations of microscale mining, diminishing with growing mine size due to the effects of economy of scale and an increasing degree of mechanization. In the two developing countries with probably the largest small-scale mining subsector, Brazil and India, the number of persons engaged in this activity is estimated to exceed 1,000,000 and 500,000 respectively.

Small mines also provide excellent opportunities for the transformation of unskilled labour into semi-skilled and skilled labour. Being essentially labour-intensive, their contribution to the process of basic skill formation can be comparatively greater than that of larger operations.

Statistics on the direct contribution of small mines to global, regional and even national mineral production are scarce and of limited reliability. The most comprehensive quantitative assessment of small-scale mining conducted on a global scale was published in 1985, reporting production estimates for 1982. The investigation, which covered the size category below 100,000 t/y run-of-mine ore production, concluded that the contribution of this segment to the global output of all major metals and industrial minerals in that year amounted to 16% in terms of gross value.

The analysis was based on an estimate of the proportion of small-scale mining of individual mineral commodities which rank most importantly in world mineral production statistics. It documented that with a number of significant commodities the share of global output produced by small-scale mining is highly impressive. With fluorspar, graphite, talc, vermiculite and pumice, this share is estimated to be 90%; with feldspar, 80%; clays, 75%; gypsum, 70%; barite, 60%; and sand and gravel, 30%. In metal mining the respective contributions are: beryllium, 100%; mercury, 90%; tungsten, 80%; chromite, 50%; and antimony, 45%.

On the basis of this detailed investigation, the contribution of the small-scale mining segment below 100,000 t/y run-of-mine production was summarized as contributing 12% to global metal production, 31% of non-metallic production, 20% of coal production, and 10% of gold and diamond output.

4.3 Environmental Issues

Mining, particularly surface mining, has come to be viewed as an environmentally destructive industry. It is apparent that the magnitude of the environmental impact of mineral extraction is related to the size of the operations and the extent and complexity of mineral processing required. Large open cast and underground metal mining operations, with their extensive associated surface facilities, are potential threats to the environment.

In the same fashion, extensive uncontrolled artisanal mining activities can cause severe environmental damage, primarily in the form of land degradation resulting from numerous excavations and waste rock piles. This is due to the widespread lack of proper environmental concern among small operators, combined with the administrative constraints associated with the implementation of environmental controls in view of the large number of operators involved and their frequently dispersed locations. In the microscale mining districts in which numerous artisanal miners operate more or less legally and regularly, the introduction and maintenance of basic environmental standards may prove particularly difficult.

While the preservation of a sound environment for future generations is a public responsibility of high priority, and is not disputed, the enforcement of stringent environmental laws has had an adverse effect on mine production costs for most operations, both large and small. Operators will, therefore, adhere to relevant regulations only if forced to do so. Reliable prevention of undesirable effects on the environment requires the preparation of environmental impact statements, pre-production environmental management plans and frequently the use of sophisticated equipment.

Individual small-scale miners will generally be unable to prepare such plans or afford to have them prepared by specialists or to acquire and operate sophisticated pollution-preventing equipment. Implementation of the necessary basic environmental standards in microscale mining districts will, therefore, require a combined approach of external assistance and control, designed to address larger entities instead of individual miners.

First, it will in most cases be inevitable that adequate pre-production environmental management plans for micro- and small-scale mining are developed by the regional or national authorities or promotional agencies related to the subsector. Secondly, it will be necessary to enhance awareness among small operators regarding their responsibility towards the natural environment. Thirdly, in artisanal/microscale mining districts, the creation of legalized organizational entities in the form of co-operatives, small operator associations or small mining enterprises will be a pre-condition for a successful environmental programme.

Moreover, in metal mining districts in which the operation of processing plants is required in order to obtain a marketable product, the installation of pollution-preventing equipment and associated facilities, such as tailings ponds, will frequently become a necessity. In such cases, concentrating mineral processing activities at a single selected site, or in a few locations, by using regional beneficiation plants appears to be the best option. In addition, special attention has to be given to the establishment of appropriate financial instruments for the accumulation of funds required for the reclamation of mined-out areas in the small-scale mining districts.

Small-scale mining legislation in the Philippines (People's Mining Act) contains a provision to this effect. It stipulates that 22·5% of governmental revenues derived from the small and artisanal operations in the form of taxes and royalties must be allocated to a special fund for rehabilitation of the affected areas. Under the new regularization programme effective in Ghana, licensed gold buyers pay the equivalent of 3% of the value of the gold purchased into the Land Reclamation Fund installed by the government.

4.4 Regional Beneficiation and Purchasing Centres

To obtain a marketable product and to eliminate the transportation of large quantities of sterile material, many minerals extracted by small-scale mining require some form of treatment and preprocessing. A decisive consideration is the concentration factor – the ratio of the quantity of run-of-mine processed ore to one unit of saleable commodity. As a general rule, the higher the concentration factor, the closer to the point of extraction the concentration process has to be conducted.

Concentration factors are comparatively low with coal and most non-metallic minerals, high with metal ores and extremely high with precious minerals. Gold ore, therefore, has to be processed right at the point of extraction, metal ores as close as possible to the mine site. To reduce the extent of beneficiation required, small miners have a tendency to pre-concentrate by selectively mining higher-grade ore, a practice which in its extreme form is wasteful and, thus, undesirable.

The availability of beneficiation facilities in the vicinity of mine sites contributes to a reduction in such high-grading practices, thereby enhancing the quantity of mineable resources and the life of the mine. In view of the limited production of small operators, the installation of captive mills at individual mine sites is usually not justified. In metal mining districts the establishment of regional beneficiation plants serving several operators has, therefore, proved successful.

Regional concentrators operated by a governmental institution have been in use in the small-scale mining segment of Mexico for more than 25 years. In Chile, Empresa Nacional de Mineria (ENAMI) operates four regional copper concentrators, treating ore purchased from numerous small and medium operators. The agency also maintains 16 mineral purchasing centres located throughout the country which serve the small miners as marketing outlets. The existence of the ore-buying stations has proved highly beneficial in stimulating small-scale mining in the country.

In Morocco, CADETAF, a governmental agency, offers similarly essential services to the artisanal miners extracting high-grade lead and zinc ore in the southern Atlas Mountains. It operates seven regional ore-purchasing centres that also provide basic materials and essential services to the miners. Purchases are made based on average monthly prices quoted at the London Metal Exchange, adjusted to account for handling, transport and treatment charges and taxes. Several mobile concentrators have recently been introduced in the artisanal gold mining segment in Ethiopia in an attempt to replace the traditional panning method.

4.5 Co-operatives and Private Sector Initiatives

While co-operatives have an image of inefficient business conduct, due to adherence to the principle of democratic decision-making, indications are that they present a favourable option in circumstances which prevent the establishment of small private enterprises.

The co-operatives established in Zimbabwe are business enterprises which operate on a self-help basis for the economic benefit of their members, who are, in effect, owner-employees pooling their resources. The conditions required for the formation and registration of a mining co-operative in Zimbabwe are as follows: there must be a group of at least 10 persons; the group must have mining rights to a mineral deposit (samples of which must be assayed by the authorities to determine the potential); in case of favourable results, the mining location must be registered with the Mining Commissioner; the deposit will then be evaluated and, if the evaluation is positive, the group can apply for registration as a mining co-operative.

Private initiatives in the form of artisanal microscale mining present the least desirable organizational choice of small-scale mining. They are usually highly inefficient and wasteful and lack safety, environmental responsibility and basic social requirements. Because of the large number of individuals and their irregular work practices, they are extremely difficult to control. Every effort should, therefore, be made to encourage the formation of suitable, larger organizational entities in artisanal mining districts, such as co-operatives, miners' associations and formal small private enterprises.

The most dynamic and generally the most appropriate type of organization in small-scale mining is the small private enterprise. The establishment of private enterprises in small-scale mining in developing countries should, therefore, be encouraged wherever possible. The removal of major barriers to entry in terms of finance and skills requirements and the creation of the appropriate monetary incentives are key prerequisites.

4.6 Financing Incentives

In spite of the comparatively modest capital requirements of small mines, securing the monetary resources necessary to finance fixed assets and working capital is a typical constraint encountered in small-scale mining. Due to severe limitations in providing an acceptable collateral, combined with the risk inherent in mining, debt financing through commercial banks is particularly difficult to obtain by the small operator. Equity financing, based on the limited personal savings of the entrepreneur and possible associates has, therefore, been the traditional method of raising capital.

Again, Zimbabwe offers an illustrative example on useful ways to alleviate the financial constraints in small-scale mining. There, a plant-hire scheme administered by the Chief Mining Engineer and executed through regional mining engineers, has been introduced by the government. Designed as a hire-to-purchase arrangement with terms between 12 and 36 months, equipment is hired out to the small miner upon payment of a small deposit. In addition, various types of loans at favourable terms funded by the government are offered to the miners.

A similar scheme of credit facilities initiated to assist the small and medium mining segment exists in Chile. As part of its services, ENAMI provides short- and medium-term loans to small mine operators, and grants emergency advances and advance payments on ore purchases. The agency also provides repayment guarantees on credits for the purchase of essential mining equipment by small miners. In Mexico, credit facilities for small and medium enterprises engaged in mining are made available through PECAM, a special complementary programme created by the government with the support of multinational lending institutions.

4.7 International and Regional Policies

4.7.1 Co-operation in marketing

The marketing of precious metals and stones produced by the artisanal mining segment is a particularly delicate matter in many developing countries due to the tendency to trade the minerals illicitly. In view of the resultant losses in governmental revenue, establishing a functioning legal mechanism of marketing outlets is a high priority objective in most countries concerned.

Empirical evidence confirms that guaranteed payment of a fair price for the product in real terms is a pre-condition to make the mechanism work. In an economic environment with an over-valued currency, this will necessitate payment of an adequate premium on commodity prices based on the official exchange rate. A number of developing countries with a significant artisanal-scale precious minerals production, including Colombia and Ghana, have successfully experimented with this instrument. Another way to create the necessary confidence in the pricing system among small producers, especially with gemstones, can be to market by public auction.

While information on international co-operation in this field is scarce, the approach followed by Zimbabwe presents an interesting option. Apart from the relevant activities of the para-statal Zimbabwe Mineral Marketing Corp., all branches of Barclays Bank in the country have been authorized to purchase gold on sight, making immediate payment to registered and regular producers. In view of the success of this system at the national level, commercial banks with an extensive network of branch offices across national borders could be considered for gold marketing on a regional scale.

4.7.2 Technical assistance

With unsafe and inefficient mining methods and work practices common characteristics of small-scale mines in developing countries, the continued provision of extensive technical assistance designed to improve working conditions must be made an essential element of sector-specific promotional programmes. The pooling of technical assistance on a regional scale, therefore, can provide a means of accelerating the process and enhancing its efficiency.

Small mines are generally not well suited for the application of highly mechanized, capital-intensive mining methods using sophisticated machinery. In an economic environment with a scarcity of capital and an abundance of cheap labour – a situation prevalent in many developing countries – the least costly method of production is one with a low-to-intermediate degree of mechanization, based predominantly on local inputs.

As a consequence, the modern mining techniques used almost exclusively in the mining industries of industrialized nations today are frequently not appropriate for small mines in developing countries.

A possible exception is open-pit mining, producing low-value bulk minerals. It follows that any transfer of technology to small-scale mining activities has to take place primarily among developing countries. Internationally active agencies, specializing in this field, such as the Intermediate Technology Development Group, based in London, and Appropriate Technology International, based in Washington, D.C., can be helpful in bringing the countries together.

International organizations, including the World Bank, other multinational lending institutions and various U.N. agencies have made significant contributions to the promotion of small-scale mining activities in a number of developing countries in the recent past. While traditional forms of financial and technical assistance provided by international agencies are continuing, initiatives increasingly focus on environmental, health and social aspects. In addition, particular attention is being given to the creation of an economic environment conducive to small-scale mining.

4.8 Human Resources Development and Technical Co-operation

In its traditional form, mining is a physically demanding and complex business requiring a unique combination of special skills for different tasks. Therefore, in most types of small-scale mining, healthy and skilled workers are a key factor for a successful operation. The level of skill required, however, varies considerably with the type and size of the operation and the commodity mined. In any given case, training requirements to a large measure depend on the existence or absence of a mining tradition. For example, implementing a small-scale underground metal mining project in a developing country with little mining tradition will necessitate extensive pre-operational training programmes.

Chile provides an example of a government-executed small-scale mining project specifically designed for human resources development. This project was funded by the U.N. Development Programme (UNDP). The objectives of the project, which focuses on two pilot areas with polymetallic mineralizations of copper, lead, zinc, gold and silver, are to provide small operators with the technical and economic knowledge required for exploration through the financial appraisal of a mineral deposit.

4.9 Fifth Tripartite Technical Meeting

The Fifth Tripartite Technical Meeting for Mines other than Coal Mines, convened by the International Labour Organization in Geneva in March 1990, confirmed the importance of small-scale mining as a significant segment of the extractive industry. The general report prepared for the meeting examined the constraints on, and the economic significance of, the industry. The report also discussed the role of governments and the private sector and addressed safety, health, environmental and training issues.

The meeting stressed the contribution of small-scale mining to the creation of employment, frequently in remote rural areas, where alternative job opportunities are scarce. While reliable statistical data could not be presented, it was indicated that in countries with an active small-scale mining segment, labour engaged in the activity can amount to 50% of total mining sector employment.

Conditions of purchase of mine output are considered an essential component of the promotion of small-scale mining. It has been observed that a difference between commodity prices paid by the government and those prevailing on the world market in excess of 5% will lead to illicit trading. The government, moreover, has an important role to play to provide an appropriate regulatory framework which generates stability in the subsector and to ensure that adequate safety and health standards are introduced and maintained.

Apart from inadequate equipment, the lack of mechanical skills and of fundamental geological, mining and processing know-how have been identified as major inhibiting factors in the successful economic development of small-scale mining. It is considered an important task of governments to provide small-scale miners with basic geological information for successful prospecting and to establish an adequate training infrastructure for technical training related to the assessment of ore grades, sound mining practices and ore-processing methods aimed at improving operational efficiency. Management training is a vital component of the training package.

Private-sector initiatives to establish small mining operations should be encouraged by the national authorities through incentives or by removing constraints and barriers to entry for entrepreneurs. The fact that the economy can benefit directly and indirectly in various ways from small-scale mineral development justifies governmental support in this process. Small mines contribute to employment, basic skill formation and infrastructure development and to public revenue in the form of taxes and royalties.

The Fifth Tripartite Meeting placed special emphasis on issues related to both the natural and the working environment of small-scale mining activities. It noted that a significant difference exists between the safety performance of large and small-scale mines. Unhealthy working conditions and unsafe mining practices, frequently observed in small mines, are made possible where government safety and health regulations are inadequate or inspection is lacking.

Amongst many examples, it was noted that the uncontrolled and inadequate use of mercury and cyanide for the recovery of gold from gold-bearing ore can have severe damaging effects on the health of the miners, the local population and the natural environment. With more than 50% of these chemical agents typically being lost in artisanal operations, the contamination of the local environment is substantial and in some cases irreversible.

Against this background, the meeting emphasized the necessity of governmental initiatives to protect the natural and the working environment in small-scale mining in developing countries. In some cases strengthening the authorities responsible for the enforcement of relevant regulations will be required if unacceptable working conditions and environmental practices are to be eliminated. The legalization of small-scale mining operations along with the provision of technical and management assistance is considered equally important in order to enable governments to implement and control basic health, safety and environmental standards.

4.10 Conclusions

Interest in small-scale mining continues to be strong in many developing countries, as demonstrated by the high level of activities in the subsector. Small-scale mining has been identified as an economic activity providing significant employment opportunities, utilizing numerous small, otherwise unexploitable mineralizations.

On the national level, a policy element designed to encourage the establishment of accountable organizational structures, such as co-operatives, small miners' associations or small private firms, will be necessary in order to eliminate environmentally and socially unacceptable practices. Accumulation of funds for land rehabilitation collected from producers is an important complementary measure. Availability of regional beneficiation facilities and purchasing centres has a strong promotional effect in small-scale metal mining districts. While co-operatives are suitable organizations, small private enterprises are preferable in view of their usually more dynamic and competitive performance. Plant-hire schemes and special loan arrangements have proved highly encouraging for small-scale mining activities, and should, therefore, be made available wherever justified by the extent of small-scale mining activities.

Regional marketing of precious minerals through commercial banks offers a simple option requiring little investment where an adequate network of branch offices exists. The pooling of technical assistance, utilizing existing regional organizations as co-ordinating mechanisms, is a promising strategy which can accelerate the promotional effect and enhance its efficiency. To ensure the use of appropriate technologies in small-scale mining, a transfer of technology will increasingly have to take place between developing countries. Human resource development is a key issue in small-scale mining, and extensive technical assistance is required in all types of small-scale mining.

CHAPTER FIVE

Mining and Environmental Taxes

Extracts taken from a paper presented by Richard Westin, University of Houston Law Center, Texas.

It is a basic concept of microeconomics that rational companies will not pay to replace common goods ('externalities') that they appropriate or degrade, such as clean air. The legal liability system is not sophisticated enough to attribute pollution to each source and to impose damages. Open access to these common goods means the full costs of pollution are not included in the cost (hence market price) of the products.

5.1 Microeconomic Analysis of Environmental Taxes

Microeconomics concentrates on small models in order to study consumption, production and pricing of goods and services. One of the general assumptions is that companies will produce goods until the marginal cost of the last good produced equals the price it can command. Microeconomic models in their simplest form involve only one producer and one consumer. Complex models add a larger number of participants and varying initial allocations of goods. The discussion that follows assumes the simplest case.

A 'demand' curve shows the quantity of a good that consumers will collectively purchase at any given price per unit. The producers have a 'supply' curve, which shows the quantity of a good that producers will collectively produce at any given price, and is derived from the producer's 'marginal cost' curve. The models that follow assume perfect competition and an implicit willingness to put some producers out of business.

5.1.1 Taxing production to cure market failure

The following graph illustrates a single firm's marginal cost curve and the demand curve for the product. The vertical axis is price per unit of output. The horizontal axis is the number of units produced.

If the company appropriated a common resource to the detriment of others, and if it were assessed damages for doing so, then its price per unit would rise. The second line is sometimes referred to as the 'social marginal cost' because it includes the full price to society. The new equilibrium would fall from Q2 to Q1 for units at the higher price.

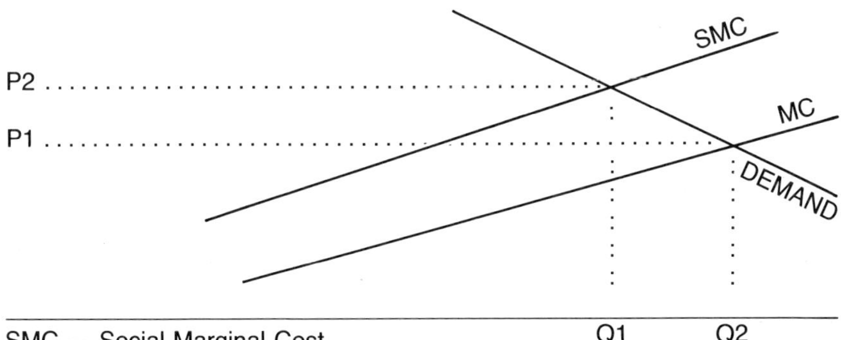

SMC — Social Marginal Cost
MC — Marginal Cost

If the remedy is to be a tax, as opposed to direct control, then the tax should be set at P2 minus P1. The tax can be set at a level that reflects damages done, or it can be set at the cost of controlling pollution. Sweden has chosen the latter course in some of its decision-making because of the difficulty of measuring damages.

There are two possibilities concerning the use of tax revenues.

First, the revenue can be dedicated to repairing the damages or, second, the revenue can be used in the most efficient manner that the government can discern. Increasing the price without repairing the damage does not in itself ensure significant environmental benefits, but it cures the 'market failure'. (Note the odd implication that equal remediation occurs if the government appropriates enough plants or product to force a price increase by reducing supply.)

While the result is an appropriate price level, it is not necessarily good environmental policy. The 'fully priced' output could still be too morally offensive to be acceptable in the minds of objective observers. Also, assuming the cost to society can be measured, the fact that a tax is imposed does not mean that the revenue will actually be transferred to the victims of the environmental damage.

5.1.2 Partially uncollectable tax

Microeconomic concepts can be used to set a tax that would induce pollution reduction in lieu of direct government intervention. The concept is to structure a system in which it is more attractive to pay to correct the problem at its source than to pay the tax.

Assume that government planners need to achieve a certain level of emissions. If they can assess the polluter's cost of decreasing emissions, the planners can derive a marginal cost curve for emissions reductions. The normal presumption is that the unit cost of abatement rises as the level of abatement increases. Planners can in theory also derive a social marginal benefit curve. The curve reflects the declining social value of eliminating emissions. For example, eliminating the last measure of pollution is likely to be very expensive to the producer and of trivial value to society.

The junction of the marginal benefit and marginal cost curves indicates where the marginal cost of abatement equals the marginal social value of abatement. This is the point at which the tax should be set. The level should be high enough to encourage substituting individual initiatives for abatement in lieu of paying a tax.

If the tax is properly set, it will not be paid until the cost of compliance exceeds the tax. The polluter will substitute technological improvements for tax payments. Beyond the point where society cares about pollution, the polluter has no incentive to avoid polluting.

5.2 Identifying Marginal Social Costs and Benefits

The foregoing economic principles depend on quantifying variables that are, in fact, elusive. The actual social cost of pollution is extremely difficult to identify. Assume the tax was set too low because the marginal costs were understated. The actual cost curve is reflected by MC and the tax will not cause the firm to avoid pollution. The government gets tax revenue.

Underestimated marginal costs

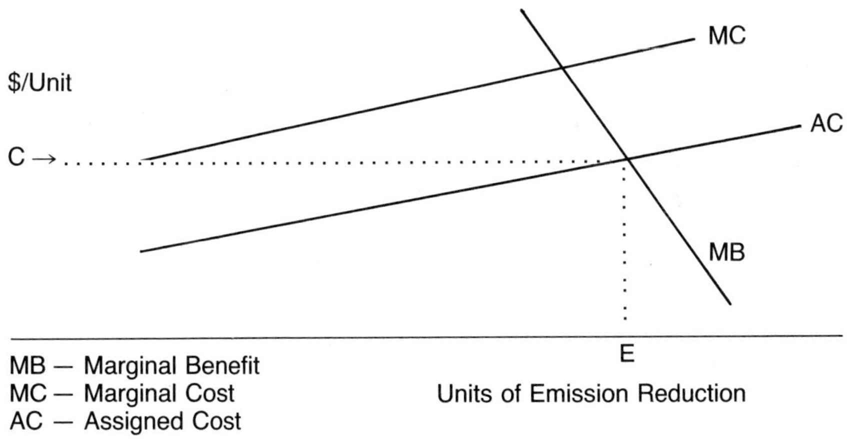

If the real cost curve is, in fact, lower than the planners believed, the tax will induce the firm to eliminate *more* emissions than is socially desirable, in the sense of allocating excessive resources to the task (i.e. the polluter spends more money on abatement than the abatement was worth).

These potentials for error encourage arguments in favour of controls, including marketable rights. Another concern that makes non-tax approaches preferable is that setting the tax may call for producers to provide marginal cost information. The tax on pollution encourages exaggerated costs.

Economic theory does not inherently reject the concept of imposing mandatory pollution standards and accompanying them with direct subsidies so that the polluter can choose the technological change that is best adapted to its circumstances. In form, a subsidy can be structured as a lavish tax credit for investments in changing technologies. However, this approach conflicts with the OECD's 'polluter pays' principle.

5.3 Accounting for the Future

Some environmental problems are of short duration and present no problems for the future. Others present exasperating difficulties. For example, some policy planners believe that in making a decision that will affect future generations, they should presume that all the affected generations vote simultaneously. How that is realistically performed is not clear, but according to this principle, environmentally-related tax policies should account for this procedure.

A related problem involves discounting cash flows. Modern theories of corporate finance evaluate projects in light of the discounted present value of cash receipts and disbursements over the period of the project. However, if discounting techniques are used in setting environmental policies, the effect can be to minimize or eliminate what are, in fact, serious long-term problems. For example, an implication of applying high discount rates is that one can virtually ignore such issues as loss of biological diversity and can tolerate global climate changes because they are so far off in the future.

The apparently correct modification is to assign lower rates to long-term environmental damage when undertaking project decisions. That obvious statement is very difficult to apply. What rate should one use? Should one distinguish among different types of damages? For example, can one treat workers' health more indifferently because the worker, being human, is going to die anyway?

5.4 Alternatives to Taxes

5.4.1 Command and control

'Eco-taxes' are not a complete solution. In fact, as demonstrated, there are serious problems in applying the convenient models of economists to this area because of the high level of uncertainty regarding marginal costs and social benefits.

Command-and-control approaches have reduced industrial pollution in free market and state-managed economic systems. Direct regulation establishes a set standard of environmental quality or pollution control that must be widely achieved. It tends to insulate the bureaucracy from corruption by restricting administrative discretion. It helps private firms in their business planning because it sets rules that rarely change and apply to all competitors.

However, direct regulation also poses problems. It is inflexible and tends to work at the expense of causing significant economic inefficiencies, inequitable competitive effects, and restricting technological development, and it can be costly to administer.

Command-and-control regulation often distorts healthy competition both within and between business sectors. Direct regulation can favour small businesses, because if there is a large population of polluters and few administrators, it is hard to force compliance. The government tends to concentrate enforcement on large polluters and to let the others go uninspected. Voluntary compliance by small polluters is not likely if they expect non-complying competitors to gain an advantage. However, direct regulation can be exploited by large firms for competitive advantage because they can support strict regulations that are more easily achieved by large-scale operations, squeezing out smaller companies and building barriers to entry of new.

Suppression of new pollution-control technology is a problem, especially if the regulation specifies the kind of equipment the polluter must use. Even if it specifies a standard of performance rather than equipment, the technical complexity of the performance standards and the sanctions for failure to meet them favour the use of known technologies over experimenting with new ones.

The worst criticism of command-and-control regulation may be its inefficiency. Direct regulation reduces the amount of a harmful substance in the environment by making each polluter use the same standard. However, costs vary from producer to producer, and economic savings would result if producers that can reduce their emissions most cheaply controlled pollution more than those who could only reduce their emissions at significant expense.

It would be expensive to administer a system of variable emission rates through direct regulation. This is where alternative strategies become interesting. One option is for the government to create a 'market' in pollution rights; another is to use fiscal measures designed to offer incentives for environmentally desirable behaviour.

5.4.2 Marketable rights

The 'bubble' theory provides that regulators can set a maximum level of pollution for a region. The marketable rights approach assumes that any surplus pollution 'rights' can be sold to those that cannot comply with the standards. Thus, firms that can reduce pollution most efficiently will be encouraged to do so (providing society with an overall efficiency) and there will be minimum government intrusion as to exactly how compliance will be accomplished.

It should be noted that, if one believes that the harm caused by pollution increases slowly as emissions rise, but that abatement leads to a sharp price jump upward at a certain abatement level, one would favour a tax, because a small mistake in tax rates would not cause the risk of a major increase in pollution or impose large unknown costs. If instead, one believes that the damage from pollution increases suddenly as one passes a certain threshold level of pollution, but that abatement costs rise gradually, one would favour marketable rights, because they would ensure that the dangerous threshold level was never reached (see S. Breyer, Regulation and its Reform: Harvard, 1982).

5.5 U.S. Initiatives

5.5.1 Feedstock taxes and superfund

The U.S. is enforcing environmental clean-up by using tax revenue when private responsible parties cannot be found. Taxes on dozens of chemical and petroleum products provide the primary public source of the Superfund to sanitize old and abandoned hazardous waste sites.

5.5.2 Tax subsidies

Section 169 of the U.S. Tax Code lets certified pollution control facilities be amortized (deducted in equal periodic amounts) over sixty months. The provision is elective, and is supposed to induce private industry to commit funds for anti-pollution facilities. The facility must not significantly change the output capacity or useful life of the taxpayer's plant, nor may it reduce the plant's operating costs. The costs must also not be repaid by the recovery of wastes.

Section 169 has been criticized on several grounds: it merely softens the blow of doing what regulators require anyway; it does not help taxpayers running tax losses; it only reaches some pollution; and it fails to encourage use of less polluting fuels. The most serious criticism is its denial in cases where it can add value or profit. (Bonds issued to finance water and air pollution-control facilities in the hands of private owners enjoy tax-exempt interest in the U.S.) It does not matter if the facility also engages in reprocessing the waste, generation of energy, or other activities ancillary to waste disposal.

The core idea is to allow a fiscal subsidy by removing a tax on income from capital to establish two forms of waste-removal facilities. This does not seem to merit much debate, but its effectiveness is questionable because many waste-removal facilities will probably be built anyway, and the subsidy may not be sufficient to stimulate increased construction of such facilities where they are needed.

5.5.3 Deducting environmental costs

This section is concerned with the tax issues that surround a mining company's duty to close mining operations safely, to remove slag and restore the earth close to its original condition. In most mining operations this will be years, if not decades, after the start of operations.

Most sophisticated taxpayers use the so-called accrual method of accounting. Under this method, a taxpayer reports income when all the events fixing the right to payment have taken place (for example, when an order of goods is shipped out) regardless of when actual payment is due or received. Likewise, a deduction is available when all the events fixing the duty to make a payment in some future year have taken place. For example, if removing the overburden above a mineral deposit triggers a duty to pay money in order to rehabilitate the land, a deduction may be charged when the overburden has been removed even though the area may not be reclaimed until well in the future.

Applying the accrual method in its pure form, the taxpayer should be entitled to claim a deduction for additions to reserves for future mine-closing expenses. Under sophisticated financial theory, applying the so-called sinking-fund approach, each annual reserve is larger than the prior reserve because the reserve is deemed to grow by the theoretical interest income that it generates as an investment.

The U.S. denial of general reserve accounting for Federal income tax purposes is partly prompted by the fact that many taxpayers manipulated the method by shrewdly creating liabilities that would not call for payment until far in the future. Nevertheless, reserve accounting is allowed in certain specified areas.

If the taxpayer cannot deduct liabilities associated with correcting their malfeasance until the correction is made, they will have no incentive to save for clean-up expenses or to pay insurance premiums to generate a cash reserve. Once clean-up becomes physically feasible, an incentive arises to clean up because doing so will produce a current tax deduction.

5.5.4 Nuclear decommissioning costs

One of the exceptions to the general U.S. rule against deductions for adding to reserves for future expenses appears in Section 468A, which lets taxpayers deduct *actual* payments made to reserve funds to pay for future costs of decommissioning nuclear power plants. The fund is carefully circumscribed so as to be usable for no other purpose and to ensure that it terminates when its purpose is achieved.

Section 468A is a sound provision. It matches income and expenditures over time, while limiting the deduction to actual payments into the fund. It is compatible with environmental goals because it encourages adequate funding for the potentially enormous costs of closing down a nuclear facility. It helps to eliminate the risk that the public might have to bear the eventual cost. This enlightened tax treatment provides a good model for funding liabilities for future disbursements to rectify unavoidable environmental damage from accepted, legitimate mining activities.

The U.S. tax law has a parallel provision for mine-closing expenses. However, it is weak and therefore undesirable. Specifically, Section 468 grants taxpayers a current deduction for the costs of reclaiming waste sites and closing mines. The amount of the deduction is calculated by a sinking fund approach. The weakness is that no actual payment into the reserve is needed in order to claim the deduction. The deduction is permissible, however, only if the reclamation involves a property reclaimed under a permit issued pursuant to specified laws. A requirement that the reserve be funded and escrowed would solve this problem by matching contributions to the escrow account with the permissible deductions.

5.5.5 Self-insurance

The U.S. tax treatment of paying for liability insurance is well settled. If the coverage has a business purpose, it is deductible as a trade or business expense under Section 162 to the extent that it covers liabilities in the current year. U.S. tax law disallows deductions for self-insurance, theorising that no risk transfer took place.

Captive insurance stands in the middle ground between commercial insurance and self-insurance. To have insurance for tax purposes, U.S. tax law requires both risk transfer and risk distribution. Thus, if an insurance company is wholly owned by another company, and the insurer insures only the parent's risk, no tax deduction can be taken for the premiums paid by the parent. Moreover, a business may not deduct alleged premiums that it disguises as a reserve which would otherwise be non-deductible.

The concept of risk transfer is that another person, and not the insured, must replace the insured as the bearer of the risk. This principle renders an alleged transfer of one's own risk to one's subsidiary defective; the risk remains with the real party in interest. However, the Tax Court has suggested that there may be risk transfer where at least 50% of the subsidiary company's premiums are received from unrelated companies. This aspect of the law is unsettled.

The risk distribution effect of insurance operates through the diversification of risks. When the law of numbers applies, unusually favourable and unusually harmful experiences tend to balance each other. In theory, unless the pool contains a sufficient diversity of risks, the law of numbers cannot apply. The Tax Court asserts that all of the risks can be of a related corporation. It has not said how many risks are enough or how one tells whether the law of numbers is operating; apparently the court would be satisfied by the application of insurance industry standards, whatever they may be.

Another logical requirement for deduction is that the insurance company has to be reasonably able to pay if a casualty occurs. If the 'insurer' is unable to pay, in substance no insurance exists. This result relates to the principle of risk distribution, for only through a pooling of premiums for diverse risks can the insuring entity acquire sufficient capital to cover a major loss.

The situation in which there is virtual certainty that the risk will eventually materialize presents a teasing problem in the insurance area. In such cases, one would assume that a portion of the premiums is equivalent to a reserve for future expenses and should not be deductible. The insurance experts evidently think in terms of uncertain risks, as opposed to certainties.

Commercial coverage is becoming increasingly scarce, just as industry needs it most. In recent years, for example, virtually no commercial insurance has been available in the U.S. for environmental liabilities associated with industrial activity. If self-insurance and reserve accounting do not afford tax deductions for coverage of environmental risks, the economics of the particular situation may drive the taxpayer to risk all by not funding the risk.

This poses difficult policy choices. One alternative is to proscribe, or sharply limit, the activity that produces the risk. Another is to facilitate self-insurance by setting clear federal standards which, if satisfied, can assure the taxpayer that a deduction will be available. Opponents of such a scheme can be expected to argue that, if industry members are not willing to engage in risk-pooling on their own, the activity should not be subsidized. That argument ignores the overall economic and social utility of the industry and overlooks the possibility of blending a variety of risks under one insurance premium.

Another approach would be to provide government insurance against environmental disasters, and to allow a tax deduction for contributions to purchase such insurance. The appeal of such an approach would be broad risk-pooling, with a greater likelihood of appropriate risk-shifting because the large number of participants should even out the actual types of liabilities that come home to roost as disasters.

5.5.6 Trust fund theory

Rather than impose a general tax, it may be preferable to identify particular activities that attract a tax, and pool the proceeds to pay an identifiable future cost. The practical effect is internationalization of cost, producing an actual fund that becomes available as needed. Apparently, the French Government has been able to get industry to embrace the concept on the theory that it is really a form of business insurance. A good U.S. example is the oil spill liability trust fund, which is imposed at a low rate on barrels of petroleum brought ashore.

5.6 Eliminating Adverse Tax Features

The following discussion relates to U.S. tax concepts. It will only be of interest if a concession agreement or tax code contains analogous provisions.

The percentage depletion allowance as it has developed in the U.S. alters the preference between extraction and use of virgin materials compared with the recycling of used materials. Because the extraction, refining and processing of virgin materials typically generates greater amounts of waste materials than recycling of used materials, the percentage depletion allowance again becomes relevant to waste management and reduction.

The percentage depletion allowance emerges from an elaborate network of rules relating to the depletion of natural resources. Despite the complexities, the system reduces to two fundamental systems of depletion, namely cost depletion and percentage depletion.

Cost depletion recovers the taxpayer's interest in an exhaustible natural resource at the rate at which it is being extracted. To illustrate cost depletion, if a taxpayer paid $10,000 for a mine containing 100,000 tonnes of coal, the taxpayer could claim a \$0·10 deduction for each tonne of coal extracted and sold. Cost depletion has engendered little audit controversy or legislative concern.

Percentage depletion permits a deduction based on a percentage of the sale price of the extracted mineral. For example, U.S. oil and gas extracted and sold by an independent producer is depletable at a rate of 15%. The controversy about percentage depletion is that it is unrelated to the taxpayer's investment in the resource.

Serious consideration should be given to determining how to balance appropriately the extraction of virgin resources as opposed to recycling used resources. The present structure of the U.S. tax law does nothing to encourage recycling. Developing countries would do well to study America's shallow approach to this issue, with a view to balancing environmental and economic interests.

A mining country can advance its environmental policies by linking tax deductions available to mineral processors with compliance to environmental norms. There are various ways to implement such a policy. At one extreme there could be a general inspection approach, under which the government advises the mining company of what standards it has to meet in a broad sense. A more rigid approach would be to dictate exactly how the process was to be undertaken, and to avoid specifying the standards. Drafting the details of such laws would not be easy, but the linkage concept provides an important influence on taxpayers' behaviour.

5.7 Workers' Health

Mining activities are notorious for their injury to human health and even life. Evaluating the 'cost' to society is very difficult, and tends to be morally uncomfortable because it calls for assigning a price to human life. Also, it may expose the particular country's failure to provide health protection of any sort; in such cases, the cost to society can be set at zero. These paradoxes can be addressed at a *minimum* level by predicting the mortality and morbidity experience of a proposed mining activity, and asking what the country will *in fact* pay to cover these risks. At a more sophisticated level, the government can ask what it would require in order not to incur these problems. Other possibilities include determining what it would cost to provide first-class workers' health coverage, and what the appropriate minimum level of company-paid life insurance cover ought to be.

The U.S. has addressed this issue in apparently only one case, namely pneumoconiosis among coal miners. The federal reaction has been a measured one. Rather than bar coal mining altogether, it has imposed safety rules and a tax ranging from 55 c to $1·10/s.ton (the lower rate for surface mining reflects the safer work standards above ground). The government has also allowed the formation of non-taxable trusts (black lung benefit trusts) that hold the proceeds of mandatory employer contributions; the trusts are used to pay health benefits to miners disabled by the disease. There is little doubt that the employers do not pay the full cost of the health hazards, but this approach is still a good one.

It is submitted that this is the correct approach for all mining activities that carry health risks, but it raises difficult issues. Should the employer bear the full health cost if the miners are unionized and can bargain for high wages? If so, what should that share be? What does one do if there are later health findings that upset the original assumptions, such as breakthroughs in medical technology, or discovery of previously unanticipated risks?

5.8 Polluter Pays Principle

The OECD seeks to ensure that no member country provides subsidies, including tax subsidies, which have the effect of diminishing the cost a polluter must bear to come into compliance with pollution-control norms.

The central concept is that the polluter should bear the cost of measures to reduce pollution to an 'acceptable state'. However, the OECD treaty only reaches a limited number of countries. Also, the principle is fairly weak. Nevertheless, as the number of OECD signatories increases, and as general sensitivity to looming environmental problems mount, the principle is likely to be of increased importance.

5.9 Profit-Sharing Structures

Whether a mining enterprise is dealing with a fully evolved tax code or an open-ended concession system, the company understands that the host country's government will exact some share of the proceeds of the mine, whether in kind or in money. The proceeds may take the form of income taxes, royalties, a profit-sharing percentage, a share of net profits or even more elaborate forms of participation.

Whatever the arrangement, there is a need for the company to earn an appropriate rate of return. With no prospect for a reasonable return, the project will not begin; with 'too large' a return, the host country will lose a source of potential revenues which may be desperately needed.

Taxes on the enterprise basically fall into two categories: those that reflect benefit the enterprise receives from government (or harm it does), and those that extract a share of the revenue of the enterprise. The negotiations should separate these issues scrupulously. Unless the host country has some other mechanism for dealing with the benefits and burdens of mining operations, the former category of taxes should always be imposed, whereas the latter are subject to seemingly endless variations.

Many countries leave the question of environmental protection and related fiscal measures to *ad hoc* negotiations. This seems unwise and, at the least, a fixed tax code should be implemented which ensures that injuries to workers' health, the cost of government administration, and environmental damages are paid for by the polluter. This is completely different from suggesting whether the revenue-sharing system should involve an income tax, royalties, or other arrangements.

5.10 Scientific Information on Environmental Risks

It is clear that there is a great deal of uncertainty about the implications of mining activities on health and the general ecology. Governments will amend their policies to some extent as new information appears. Because it is costly to generate such data, and it is not always disseminated broadly, mining countries should consider pooling resources to keep abreast of such developments. The U.N. is, of course, helpful in this regard.

The study of environmental implications of mining activities is a relatively new field. To the extent tax laws grow out of assumptions about environmental effects, when the scientific information changes, so should the tax laws. This fact should be made clear to governmental policy makers and politicians. One implication is that the distinction between tax laws designed to get revenue and those designed to advance environmental values should be kept alive, and politicians should not be permitted to rely on such taxes if the scientific basis for the concern that drives the tax is unstable.

5.11 Recommendations

There needs to be a 'cradle to grave' approach to environmental problems; it is not enough to design a creative response, there must be follow-through. If a tax on pollution turns out to be too high or too low, the results will be perverse. If they are too low, the firm will not change its behaviour. If they are too high, there will be 'overcompliance', with the resultant costs in excess of what society reasonably requires of the polluter.

This calls for sensitively monitoring the impact of the tax. Failure to verify the impact of environmental taxes has been documented by the OECD. The failure to assign economists and statisticians to follow up on environmental tax remedies is a common problem that must be corrected. In addition, toxicity assumptions should be subjected to periodic modification, upward or downward, based on later scientific conclusions.

There should be a flexible approach to environmental controls, for which there are three primary mechanisms: regulatory intervention, marketable rights to pollute, and taxes. No one approach is always best. Policy makers should not subscribe dogmatically to any one of the three as the invariable solution.

Certain taxes seem irrefutably appropriate. One is a tax on output that is designed to fund the cost of worker morbidity resulting from mining activities. The U.S. black lung tax is a good example because it is in proportion to extraction. The revenues are set aside to fund predictable medical needs that would otherwise be borne by workers and their families.

A second tax-related concept with great merit is a deduction for contributions to reserves for mine closing and land reclamation costs. Assuming the law or concession agreement imposes such an obligation, the appropriate pattern is to use a sinking fund approach under which contributions are made in cash to an actual fund that grows at interest (taxable or not) in a manner designed to ensure that the funds are there when closing or reclamation is required.

Co-operative undertakings are important. Multinational corporations communicate with each other about their industries via trade magazines, industry associations and informal networks. Developing countries can minimize economic waste by undertaking shared studies, and disseminating the information. This should permit countries to obtain the best advice on a steady basis.

Tax stability is important. All private investors seek some particular level of economic return from their investments. To the extent the return is affected by political instability or other uncertainties, the return they demand will rise, reducing the flow of capital. To minimize the required return, countries should reduce political, economic and legal instability. The latter can be helped by assurances that there will be no change in the law for fixed periods of time.

CHAPTER SIX

Economic Aspects of Environmental Protection

Extracts from a paper presented by Olle Östensson, Senior Economist, Commodities Division, U.N. Conference on Trade and Development, Genève, Switzerland.

6.1 Introduction

In the theoretical literature on international trade the environment is most often treated as a 'third' factor of production, in addition to labour and capital. A country is thought to have an environmental abundance if it has a large 'assimilative capacity', that is, a greater ability to tolerate pollutants and other environmental damage. It should be noted, however, that assimilative capacity is extremely difficult to measure in any unambiguous way, since it includes several components measured — if they can at all be measured — in different units. Furthermore, if assimilative capacity is to be used in economic analysis it has to represent not simply the physical capacity of air, water and land to tolerate pollutants but rather the level of environmental damage that society is willing to tolerate.

Setting aside these difficulties, it would be expected that if countries have identical production, pollution and abatement functions for a particular item, then the country with the higher assimilative capacity would specialize more in pollution-intensive items. Unilateral introduction of environmental regulations by this country would impose environmental control costs on its producers of the item, thus eroding any price advantage they have relative to producers in other countries.

A large number of studies have attempted to estimate the impact of environmental control costs on price and output in a range of industries in industrialized countries and on the trade of these countries. Although the methodologies used vary, a general conclusion is that total environmental control costs are lower than expected. Reductions in output caused by these costs are also small except for some sectors, including mining and metal production, where they can be significant. Finally, there is little evidence of any significant overall impact on trade at this stage.

However, it should be noted that virtually all of the studies relate to possible effects of environmental regulations in the U.S., and that the results may not lend themselves to generalization. Furthermore, they do not attempt to identify the countries that may have replaced the U.S. producers. Therefore, the studies provide no support for the argument that developing countries may decide to use relatively 'soft' environmental regulations to attract investment and provide a competitive edge, that is, become 'pollution havens'. Neither do they refute it.

There are, however, other reasons to doubt the validity of the argument. First, it may be mistaken to believe that developing countries *per se* have a higher assimilative capacity than developed countries. Many would argue that the physical environment in developing countries is no more able to tolerate pollution than that of developed countries and, in fact, may often be less able to do so because of climatic conditions. Furthermore, as already mentioned, assimilative capacity has to represent the preparedness of society to tolerate environmental damage. There does not seem to be a significant difference in public opinion on matters concerning the environment between developed and developing countries. Many developing countries have either reinforced their environmental legislation recently or are in the process of doing so.

In many cases, the reinforcement of environmental policies entails a willingness to sacrifice potential production. For example, development of the Gandhamardhan bauxite deposit in Orissa,

India, was halted for environmental reasons, and while the project has subsequently been modified it is still not certain whether it will go ahead, although it could provide a source of raw material for the state-owned alumina refinery in Phutakapahar, Madhya Pradesh. Finally, although differences in environmental control costs may be identifiable, they may not have been large enough to outweigh other factors affecting competitiveness.

At the same time, it is clear that in many developing countries, efforts to protect the natural environment may compete with, and may sometimes have to assume second place to, other concerns which are more pressing, chief among these being the need to reduce poverty.

6.2 Costs of Environmental Protection

The estimates of costs of environmental protection measures in mining and metallurgy that have been published over the past several years mainly relate to the situation in industrialized countries. The financial impact may well be different in developing countries, in particular when consequences for all factors of production are taken into account. Nevertheless, these estimates provide useful indications of the magnitude of the costs concerned.

It should be noted that the estimates have tended to increase over time in line with the strengthening of environmental regulations in industrialized countries. A study by Charles Pearson from 1976 quoted various estimates of environmental control costs in the U.S. For primary metals, the estimated costs, including capital charges, were generally around 2% of the value of shipments. The estimates indicated that costs were expected to increase from the early 1970s to 1980. Another study, on the Federal Republic of Germany by Walter Lötz, gives estimates of direct costs of the operation of installations for environmental protection as a percentage of turnover. For mining, the percentage was 0·5% in 1979, for iron and steel production it was 2·6%, and for nonferrous metals it was 3·1%. Environmental investment as a proportion of total investment in the three industries from 1975 to 1979 were estimated to be 2·9%, 17·3% and 20·0% respectively.

More recent estimates, which take into account subsequently implemented or planned changes in regulations, are generally higher. For instance, the average cost to the U.S. copper industry of meeting the requirements, regarding sulphur dioxide emissions, of the Clean Air Act has been estimated to be $0·08/lb of copper in 1987, corresponding to 45% of average total smelting costs, or almost 10% of the average price of refined copper in 1987.

The direct costs of complying with regulations in the primary lead industry in the U.S. have been estimated to be between 2·5 and 3·2 c/lb of refined lead, or 9 to 12% of the average price of refined lead in 1987, with the range reflecting the difference in operating costs between rehabilitated plants and newly constructed plants with cost saving technology. When capital charges are taken into account, the total cost reaches 12 to 15% of net costs.

Some qualifications have to be attached to these cost estimates, or, for that matter, any estimates of the costs of compliance with environmental regulations. Firstly, the figures generally refer to retrofitting of existing smelters, not new projects. In general, it can be assumed to be less costly to undertake the investments necessary for compliance when a new installation is built than to make changes to existing plants. Secondly, estimates usually cannot take into account possible positive effects on operating costs resulting from the introduction of new environment-friendly technology.

If only mining operations, excluding the metallurgical stage of production, are considered separately, the distinction between new investments and rehabilitation of existing installations becomes even sharper. The most important detrimental effects on the environment from the mining operations result from non-rehabilitation of the mining site and leakage from tailings dumps and ponds into water courses and the groundwater. In both cases, appropriate measures have to be taken at the start of the operations (topsoil has to be removed and stored, tailings heaps have to be located and insulated so as to minimize the risk of leaking of waste products, etc.). With regard to other negative effects of mining, such as dust, noise, etc., the same principle applies. Once the mine is in operation, however, remedial action to rehabilitate a mine site becomes costly, sometimes prohibitively expensive.

While the estimates vary, it appears that environmental control costs may be a significant component of production costs, if only at the metallurgical stage and for existing installations. It has not been possible, however, to identify any competitive advantage accruing to developing country producers from differences in environmental control costs. A more definitive conclusion would require a detailed study of production costs in all the countries concerned. The information needed for such a study is not available, at least not publicly.

One factor that argues against differences in environmental control costs being a significant advantage for developing countries, leading to enhanced competitiveness, is the relatively advanced age of the capital equipment in most developing countries. Thus, even if environmental control costs are lower, this advantage would tend to be outweighed by higher operating costs, which furthermore have tended to increase quickly in developing countries during the last decade, as low prices have not generated the necessary cash flow to finance upgrading of equipment or, in some cases, even to cover maintenance.

6.3 Trends in Technology Change

Unlike many other industries, the diffusion of new technology in the mining and metallurgical industry is relatively fast. By contrast, the general pace of technological development has, with few exceptions, been relatively slow and characterized by incremental innovation rather than revolutionary change. Furthermore, fast technological diffusion applies to new projects. In the case of existing installations, changes in technology will normally only be introduced in connection with major additions to capacity.

Consequently, while new projects will generally use 'state of the art' technology, they will coexist with other installations using technology of a much earlier vintage. The result is a high degree of standardization of technology. A recent report to UNCTAD by a consultant notes that: "In the mining and metallurgical industries many of the most important technological advances are developed and introduced by equipment manufacturers. Technology transfer within the industry in this way has been very rapid . . . This has led to a situation where mining and metallurgical technologies for most minerals and metals are relatively stable, standardized and widely diffused."

Another reason for the standardization of technology in the mineral sector is the desire of financial institutions for assurance that projects use proven technology, preferably from well known suppliers. Furthermore, since in many cases new projects are conceived as joint ventures, the wishes of foreign partners to use familiar technology is often a decisive factor.

Technology for environmental protection is also likely to be diffused rapidly to new projects and to conform to developed country standards. There are at least three reasons for this. First, the standardization of technology makes it simpler for the technology supplier to put in technology that meets standards in developed countries rather than try to achieve savings by tailoring this particular part of the production process to local regulations. Second, major mining companies investing in developing countries are often unwilling to put their investment at risk by exploiting a softness in local environmental standards. Regulations may very well change in the future, more stringent standards may be introduced, and the measures that could have been taken at low cost at the time of the initial investment may have to be carried out at much higher cost later. Third, government-backed financial institutions generally make their support to mining and metallurgy projects conditional on the preparation of environmental impact studies and the inclusion of environmental control technology.

The preceding is likely to apply, however, only to investments in large new mining projects. For medium- and small-scale mining projects, external finance and the necessary technology are usually not available.

6.4 Future Developments in Developing Countries

An increasing proportion of new investment in mining and metal production over the next several years is likely to take place in developing countries for a number of reasons.

First, there has been a shift in developing countries in general in their attitude to foreign investment. While governments are obviously still anxious to ensure that investment takes place under the best possible conditions and that the government obtains a fair share of the mineral rent, they are now actively seeking foreign investment as a source of finance. Only in a few developing countries will it be possible to finance major projects on the domestic capital market.

Second, major mining companies, for their part, appear to be beginning to take a more positive view of investment in developing countries than before. This is of course partly a result of the changed attitude in developing countries.

Third, major unexploited deposits are scarce in most developed countries as these areas are relatively well explored. In developing countries, on the other hand, there is often a backlog of projects that have not been developed for lack of financing, and exploration work has been much less intensive.

In should be noted, however, that the economic liberalization process in the Commonwealth of Independent States and the countries of Eastern Europe may open up alternative opportunities for investors in the mineral sector. It remains to be seen to what extent this may affect investment in developing countries.

As indicated, investment in new projects is likely to take place using technology that is standardized and the most modern available. However, the situation as regards already existing installations is more complicated. In most developing countries, the older mines and metallurgical plants use technology which has sometimes not been upgraded since the time of the initial investment. To rehabilitate these installations in order to make them environmentally acceptable would entail large expenditures.

The situation of the medium- and small-scale mining industry in developing countries also raises problems with regard to environmental protection. Medium size mines are often owned by domestic companies with a small capital base and little access to external technology. They are unlikely to use technology that limits environmental damage. Artisanal mining can also have significant negative effects on the environment, as is shown by recent experience in Brazil, the Philippines and several African countries.

Investment on protecting the environment in the developed market economy countries has been facilitated through the use of fiscal incentives such as tax deductions or government subsidies. While the policy mix has differed from country to country, it can be said generally that the public has borne a significant part of the cost of rehabilitating polluting industries through the government budget. Although OECD countries have agreed to the 'polluter pays' principle regarding the financing of environmental control costs, this has had limited practical consequence and negative incentives have been little used. However, this principle is now beginning to be taken more seriously.

Developing countries can now be said to be faced with the need to undertake the same measures in a much shorter time. As is often the case, these countries have to leapfrog technologies and bring their industries in line with standards evolved by developed countries. The main problem is likely to concern existing installations and the medium- and small-scale mining industry. For most developing countries, government subsidies for the rehabilitation of these operations are unlikely. On the other hand, in the absence of radically improved minerals and metals prices, forcing the owners of the installations concerned to undertake the necessary investments is also unlikely to succeed. Moreover, the enforcement of environmental regulations through negative economic incentives in the medium and, in particular, the small-scale mining industry is administratively difficult, and runs the risk of unacceptable consequences for employment in the short term.

A particular problem is posed by negative effects on the environment resulting from the consumption, rather than the production, of minerals and metals. While such concerns are likely to be more important for agricultural commodities than for minerals, they can lead to increases in processing costs for some minerals if a specific element has to be removed. One interesting example of how the necessary changes in the production process may be financed is the decision by the Commission of the European Economic Community to provide finance totalling ECU 15 million as part of SYSMIN (the instrument of the ACP-EEC Convention allowing for support of export earnings in the mining sector) to Senegal for its phosphate industry. Because of new environmental norms which concern the cadmium contents of fertilizers, the Senegalese industry will have to make large investments. Intervention by SYSMIN is aimed at supporting efforts made by the three Senegalese companies concerned to find a solution to the cadmium problem.

6.5 Conclusions

It would appear that the effects of environmental measures on the international competitiveness of the mining industry of developing countries are likely to be limited. New major installations are expected to be competitive both in the economic and the environmental sense. Old installations will probably not be burdened by the full cost of rehabilitation at once. It is anticipated that most governments will opt for a policy of gradual change, and are unlikely to be able to afford the rehabilitation of old installations. Neither positive incentives in the form of tax reductions or subsidies nor negative incentives in the form of fees can be realistically contemplated in most of these countries.

Consequently, rehabilitation of existing installations will take place gradually and will not significantly undermine the competitiveness of the operations concerned. While any earlier competitive advantage that might result from lower environmental control costs would be eliminated, this would in most cases be outweighed by increases in productivity due to new capital investment and changes in production technology. Similar policies are likely to be followed for the medium- and small-scale mining industry.

Developing countries will, however, require external financial and technical assistance in order to undertake the investments necessary to rehabilitate older installations and to introduce changes in the medium- and small-scale mining sector. In the absence of the necessary assistance, it will be possible to implement the measures needed to reduce damage to the environment only over a considerable period of time, and the resulting differences in the application of environmental regulations may raise trade-related problems.

The need for technical assistance also raises the issue of possible international environmental standards for the mining and metallurgical industries. Such standards would probably allow better focusing of assistance efforts and may also contribute to an improved perception of the particular problems of the industry on the part of policy makers and the general public. The standards would have to take into account the different ecological characteristics of individual countries. There is also the question of who should be charged with the development of the standards. Should individual commodity organizations, such as study groups, have this responsibility or should it be entrusted to an organization with a broader mandate, such as UNEP, in order to ensure comparability of standards for different minerals?

CHAPTER SEVEN

Environmental Management

Extracts from a paper presented by Alyson Warhurst, Science Policy Research Unit, Sussex University, U.K.

7.1 Background

Regulatory frameworks are growing in complexity, particularly in the industrialized countries. However, skills and knowledge on the part of regulators often fail to match regulatory aims and there still exists inadequate scientific knowledge of toxicity pathways and technology solutions to some clean-up problems. Moreoever, regulations alone will not solve the environmental degradation problems of the developing countries, principally because production efficiency is a precursor to achieving environmental efficiency. Enforcing regulation requires skills and resources beyond the budget and capabilities of many developing countries. It is also impossible to monitor the many remote small- and medium-scale mines. Fines are a fraction of costs of abatement technology and only payable if the guilty parties are detected and convicted. Existing 'command and control' type regulation tends to deal with the symptoms of environmental mismanagement (i.e. pollution) rather than its causes (poverty, lack of access to technology and capital, etc.).

Meanwhile, the far-sighted international companies are changing the parameters of mineral production. Contrary to the belief that a trade-off is required between corporate profit and environmental costs, they are designing equipment which will improve both the economic and environmental efficiency. Technical change is reducing both production and environmental costs. The development of new lower cost, less hazardous alternatives to smelting (for example leaching) may also be to the advantage of the developing countries. But such investments could herald another wave of technological dependency – the swapping of one social cost for another to the benefit of foreign capital.

7.2 Inadequacy of Environmental Regulation

There are a range of reasons as to why environmental regulations, particularly those of the 'command and control' and incremental 'paper tiger' nature, do not improve environmental management in developing countries. These are discussed below.

One of the major reasons why environmental regulation alone will not solve environmental problems in developing countries is due to the latter's inherent problems of production inefficiency. Inefficient operating regimes, excess capacity, breakdowns and poor management procedures all contribute to worsen the pollutant nature of effluents and emissions. Moreoever, such inefficiencies make it very unlikely that environmental controls will be incorporated effectively. Notwithstanding those obstacles, environmental regulations in developing countries tend not to work.

Production inefficiency is endemic amongst many mining enterprises in the Third World, and problems of environmental degradation cannot be viewed independently of it. In the industrialized countries, new concentrators and roasting plants tend to be totally computerized, and automatic ore assaying techniques give an extremely accurate picture of the chemical composition of the ore feed. This degree of control enables an accurate prediction and monitoring of emissions. However, where these controls are missing and, in particular, where ore feeds are of variable composition – in terms of the sulphur, lead and arsenic content – emissions can also vary with regard to their pollutant content.

Environmental regulations tend to be of the blanket type which specify maximum levels of emitted substances, minimum levels of environmental quality, and best available technology standards. They do not tend to reflect the propensity of a particular operation to pollute, which in part depends on local site-specific conditions (geology, geography and climate) as well as economic, infrastructure and technology-related constraints. In a desert, tailing dams need not be as highly specified as in rainy climates; dust regulations may need to vary depending on topography and prevalent winds; the substrata of leach ponds might need to be of different composition, strength and depth, depending on local geology (for example the existence of an impermeable level of clay).

Since developing country regulations are often copied from the statute book of the industrialized countries, they may not be appropriate for the site-specific characteristics of mines in either tropical regions or deserts. This may result in unnecessary and costly adaptations on the one hand, or the lack of necessary controls on the other.

Command and control regulations require intensive monitoring to ensure that they are enforced. However, the small- and medium-scale mining sector accounts for over 25% of mineral production in many developing countries. These mines are often situated in remote locations and are almost impossible to monitor systematically. Indeed, such monitoring requires skills and human resources far beyond the technological and managerial capabilities of many developing countries.

Moreover, the enforcement of command and control regulations depends on a system which admonishes with imprisonment and fines. This in turn requires a legal structure and judicial system far beyond the capacity of most developing countries. Compliance is also limited since fines are generally a fraction of the costs involved in remedial treatment and abatement technology. They are also only payable if the polluter is detected and convicted.

Environmetal regulations, since they are frequently devised by independent government departments, often contradict other economic and industrial policies. For example, several countries with tropical forests have recently introduced policies aimed at their conservation. At the same time, countries such as Brazil, Ecuador and Columbia have parallel economic policies to promote industrial investment, especially by foreign firms, in these remote areas.

Another recently discussed example of potentially contradictory policy issues revolves around the international Basel convention which restricts the inter-country transportation of toxic wastes. Since certain scrap metals fit theoretically into this category on account of their heavy metal content, this would undoubtedly restrict trade in scrap metal and metal recycling. This contradicts the intentions of European and American Governments to encourage the recycling of metal-containing materials at the expense of new primary production.

Command and control regulation tends to identify and deal with symptoms (i.e. pollution) rather than causes (poor environmental management practices). It is also incremental in nature. Therefore, there is a tendency for it to emphasize end-of-pipe, add-on and capital-intensive solutions (e.g. smelter scrubbers, mine water treatment plants, dust precipitators, etc.) for existing technology and work practices rather than to promote alternative environmental management systems or technological innovation. Regulation may also, to a certain extent, pre-suppose a static technology.

If regulation is incremental, technical change may be incremental, involving the adding-on of numerous new controls at relatively greater cost and with more overall resultant degradation than if a new, more radical change had been introduced in the first place. In turn, this regulatory approach may get a less co-operative response from industry, which sees the rules always changing and its costs increasing.

7.3 Environmentally Sound Alternatives: Threat or Opportunity?

After a period of using rather static technology, the mining and mineral processing industry is going through a phase of technical change as the more dynamic companies are innovating to escape economic as well as environmental constraints. Rapidly evolving environmental regulatory frameworks in the industrialized countries and the prospects of their application in the developing countries are reinforcing this trend. Changed technological and environmental behaviour in this context is particularly evident amongst the larger U.S. and Canadian mining companies, but is becoming increasingly apparent in developing country-based companies. However, State-owned enterprises and small-scale gold and tin alluvial mining groups in the developing countries continue (with some exceptions) to be both economically and technically constrained.

It is somewhat inevitable that only those companies which are dynamic will be in a position to invest in the necessary research and development required to introduce environmentally sound alternatives. Nonetheless, after a long period of only conservative and incremental technical change, quite radical alternative process routes for mineral production are being developed. These are emerging as being substantially more economically efficient as well as environmentally less hazardous than existing systems. Furthermore, these companies are beginning to sell those technologies.

The figure below illustrates the relationship between production costs per unit of output and evironmental costs. New technology is being developed which lowers both environmental and economic costs and pushes forward the frontier of leading technology. The technological frontier is a generalised band within which most mining companies operate, to a greater or lesser extent absorbing the environmental costs resulting from their activity. Companies with ongoing operations, high sunken costs in existing facilities, and less dynamic in their technological behaviour would tend to experience environmental regulatory pressures as a cost burden.

Other companies are innovating by building a new generation of technology, and protecting themselves from having to undertake costly add-on incremental technical change at later stages. Examples included Homestake's pressure leach system, Inco's nickel flash smelter and Sao Bento's gold pressure oxidation process.

This trend in technical change may be to the benefit of developing

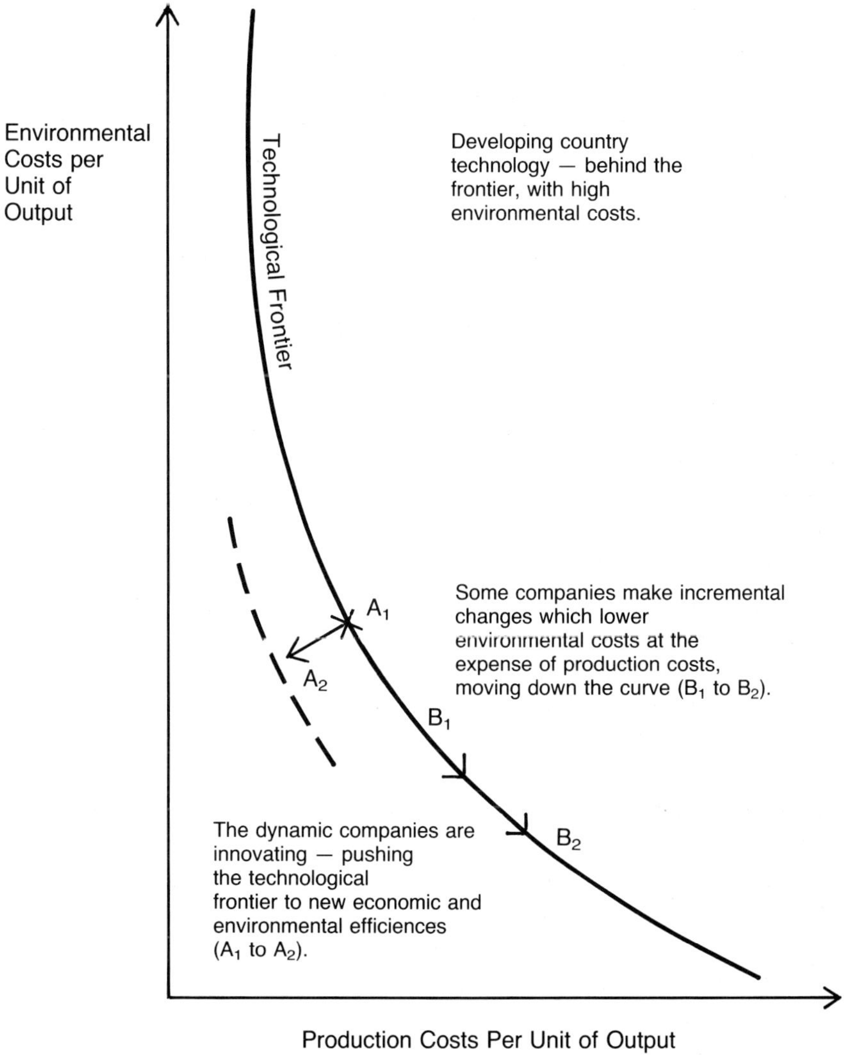

countries, in that it may enable them to reduce the trade-off between higher environmental costs and lower production costs. This may mean that, at least in the case of new mineral projects, there may be a wider range of more environmentally sound and economically efficient technology readily accessible. For example, new processing techniques mean that saleable metal can now be produced at the mine site – while in conventional process-routes the ore may have been exported to foreign smelters (with consequent loss of by-products and charges for the treatment of pollutant elements).

However, this new prospect of environmental security may have its own costs which require careful analysis. For example, could such new 'environmentally friendly' investments herald a new wave of technological dependency – whereby one social cost (environmental degradation) is exchanged for another (indebtedness, bankruptcy of local equipment suppliers and engineering firms, etc.), reinforced by aid conditionality? Or will smelting and concentrate-leaching alternatives be developed by international companies, adapted to new environmental regulations in the industrialized countries, which will oblige developing countries for economic and environmental reasons to export only semi-processed minerals?

7.4 Summary

The economic future and environmental sustainability of mineral production in developing countries poses at least four major policy challenges.

First, whether the new 'Environmental Imperative' offers an economic opportunity or threat to developing countries will depend upon how companies plan and negotiate their own participation and environmental interests in new mining projects. However, the industrial strategy and environmental policy of these developing countries need to be developed in parallel, shaped by the economic, social and political processes. These will determine the need for environmental regulation and condition company responses to it.

Second, achieving and sustaining production efficiency in the developing country context will be an important prerequisite to achieving environmental efficiency. The achievement of improved production efficiency and environmental management will, in turn, be dependent upon the extent to which technology transfer and training clauses are incorporated into joint ventures, investment arrangements or loan conditions. It is important that engineering and managerial capabilities in the developing countries are enhanced.

Third, environmental degradation from mining and mineral processing activities could be reduced and prevented through designing country-specific flexible regulatory frameworks. These regulations should deal with the potential causes of degradation, not just the symptoms of pollution, by stimulating the development and diffusion of well-adapted innovative site-specific technological solutions, and 'best-practices' in environmental management based on an informed understanding of company behaviour in different economic and regulatory contexts.

Fourth, whether a mineral-producing economy can seize these potential environmentally sound development opportunities, or suffers a debilitating round of technological dependency, will depend on its economic and historical conditions, its policy framework and the strategies of the mining companies investing there.

CHAPTER EIGHT

Socio-Cultural Impacts

Based on a paper presented by Dr Allen Clark, Director, Minerals Policy Program, East-West Center, Hawaii.

My basic premise and experience is that the continuing social-cultural impact from mining operations continues largely unabated because of the following:

1. The nature of the impacts is poorly understood and documented.
2. Social-cultural impacts are looked upon as local in importance.
3. Basic cultural and social value systems are irreversibly damaged and the process is self-destructing.

Before discussing the nature of social-cultural impacts it is first advisable to define the two terms: social impacts are those that affect interaction between people or groups; and cultural impacts are those that affect the basic beliefs and values of the people or groups.

Social-cultural impacts from mining vary in scope, intensity and importance relative to the scale of mining that is undertaken. As an example, large-scale mining activities generally result in the social-cultural problems and impacts often associated with medium-sized cities. In particular, such activities tend to attract a migratory and diverse population of people who develop a new but reasonably stable set of social-cultural values different from those of the surrounding areas and the areas from which they have migrated. Such developments have profound social and cultural impacts on existing populations. Well-known examples include OK Tedi in Papua New Guinea, Ertzberg in Indonesia and the Trombetas bauxite and Carajas iron mines of Brazil.

Medium- to small-scale mines have similar social-cultural impacts to large mines but in many cases vary significantly, particularly in terms of their impact on the existing populations. In particular, medium- to small-scale mines incorporate the existing population as employees or suppliers to the mine, thereby focusing the impact more directly on the indigenous peoples.

Artisanal mining represents an entirely different type of mining with quite variable social-cultural impacts. Specifically the miners are culturally, socially and ethnically diverse, highly migratory and unstable as a population, and often in conflict with indigenous land owners and peoples.

Although the social-cultural impacts of mining vary greatly, there are certain common factors that exist. Among the most significant are that:

1. Social and cultural impacts are largely irreversible.
2. The largest changes are to traditional values.
3. The transition to a cash economy is the largest catalyst to change.
4. Impacts arise because the process of change is rapid (Push) but the capacity to change is slow (Pull).
5. Greatest conflicts arise between the tribalists and the developmentalists.
6. Social-cultural impacts have local, regional and national implications which are largely not understood by government.

As examples of the type of social-cultural impacts defined above, it is pertinent to examine the specific issue of social-cultural impacts of mining activities on indigenous people in remote areas. As mining activities move into new frontier areas, they are coming into contact with indigenous peoples with strongly held cultural beliefs and strong social ties and relationships. Bougainville and the Panguna mine and the recently developed Misima mine, all in P.N.G., are excellent examples. The peoples of these areas held strong cultural values, in particular that land and property is:

1. Common.
2. Cannot be sold.
3. The environment should be protected.

Some of the types of social-cultural impacts caused by mining which are not well understood include the hiring of young people (under 40 years of age) within indigenous cultures that are largely governed on the basis of seniority. The result is that a young, wealthy and competitive group is developed which challenges this cultural trait. Moreover, the cash economies arising from mining revenues create a division of haves and have-nots in societies whose culture is based on shared wealth or disdain for consumptive wealth.

Other social-cultural impacts caused by mining include changes to the peoples' cultural concept of 'shared' land. This is distorted when areas are relinquished for mining as ownership and distribution of revenue become key issues within the community. Additionally, cultural precepts of the environment are similarly distorted with the environment becoming secondary to the mining operations. This becomes a larger issue with time as the full impact of mining on the environment becomes both known and visible. In summary, the land is irreversibly changed by mining, and hence so are those that occupied it previously.

The above represent only a very minor portion of potential problems and areas of dramatic social-cultural change within indigenous societies but show that such changes impact the very essence of such societies. Equally important is the recognition that such changes occur throughout the life cycle of the mine. As examples, research by the East-West Center has shown that:

1. Early stages of exploration and development result in the most conflict between the mine and the indigenous people.
2. The mining phase is normally one of reconciliation between the mine and the people but one of greatest disruption within the social-cultural system of the indigenous peoples.
3. Long-term mining results in renewed conflict between the mine and indigenous people, particularly the younger generation (who may not benefit from earlier royalties and payments). These

people may feel disenfranchised — largely reflecting the breakdown of the cultural peer system.
4. Exhaustion and abandonment of the mining venture is the most traumatic phase, leading to major social conflict within the area and a complete reassessment of cultural values.

Basically, this system of actions and reactions results in a social-cultural breakdown consisting of five specific elements (Filer 1990):
Delineation. Boundary of the mining activity defines who will be the haves and the have-nots, resulting from sharing of royalties, compensation, etc.
Distribution. Physical allocation of the revenues and establishment of long-term patterns of wealth.
Stratification. Based on 1 and 2, a cultural stratification emerges which may alter or reinforce cultural values and social organization.
Inheritance. Transfer of wealth within the culture normally resulting in a dilution of wealth and an altering of the stratification.
Succession. The sequence of distribution will change with time and will most likely result in a cultural change based on acquired wealth.

The majority of the above considerations apply to large-, medium-and small-scale mines but are less applicable to artisanal mining, which represents a somewhat special case. Specifically, in the last ten years there has been a marked increase in artisanal mining in the Philippines, Brazil, Indonesia, Ghana, China, Laos, Costa Rica and Burma.

There have been very large social-cultural problems in these areas. Impacts have included the following:

1. Health and Safety. Mercury pollution, mine cave-ins, mudslides and sanitation.
2. Child Labour. With attendant lack of education and rights.
3. Landless People. Highly transitory people unable to return to other activities.

A particularly critical issue in the longer term impact of artisanal mining is the relationship with other sectors once the activities cease (as a result of declining gold prices, exhaustion of resources or governmental controls). In particular, how will these people be absorbed in the formal or informal economy. Overall the problem, which is now a major social-cultural challenge, may become even more critical in the future as either artisanal mining becomes an even larger activity than it is at the present, or the activity shrinks considerably, with a massive displacement of people.

The spectrum of problems, and needs, regarding social-cultural impacts of mining is very large, but certain critical elements must be considered. These include but are not limited to the following:

1. Central to dealing with social-cultural impacts is the need to include those affected in all discussions and planning pertaining to the project — from initiation to final reclamation.
2. New approaches and methodologies for assessing and dealing with social-cultural impacts must be developed. These activities should be supported by the major international funding agencies and must accommodate the inputs of social scientists.
3. Base line cultural studies are a necessity and must be undertaken and supported by government with respect to all potential cultures impacted by planned developments.
4. Social-cultural impact statements must be conducted by both industry and government, including continual follow-on analysis as mining progresses.
5. Social-cultural impact studies should be collected and subsequently made available, with the U.N. taking a lead position.
6. Training in cultural and social assessments and subsequent implementation and actions should be a major emphasis of development assistance agencies and national governments.

The social-cultural impacts of mining have in most cases been inadequately studied. The resultant problems represent one of the most serious and continuing problems for government and industry in the mining sector. It is apparent that this issue can no longer be avoided. If it is avoided, it will be at the peril of local and/or national stability.

CHAPTER NINE

Towards Environmental Standards

Extracts from a paper presented by Ayesha Dias, UN/DTCD, New York.

9.1 Introduction

The mining sector is of utmost importance to developing countries for a variety of reasons: as a significant source of export earnings, as contributor of a base for industrialization and as an employer. However, as the UNEP's Technical Guide on the Environmental Aspects of Mining frankly states, "the development of a mining project necessarily modifies the natural environment of the area where it is located". Indeed, the environmental impacts of exploration, extraction and processing of mineral resources can be considerable, and the need for environmental guidelines can scarcely be disputed.

However, so far as mining in developing countries is concerned, such environmental guidelines need to be clear and effective while at the same time not acting as an impediment to much-needed investment. Moreover, there is an important role to be played by multilateral and bilateral development assistance in sharing some of the substantial costs involved in coping with the environmental impacts of development.

The process of establishing international environmental principles and guidelines and national regulatory standards regarding mining in developing countries will necessitate a complex series of interactions between several important actors. These will include representatives of developing country governments, the mining industry, multilateral and bilateral development agencies, associations of mine workers, environmental non-governmental organizations, and communities affected by the mining activities. UN/DTCD seems ideally placed to initiate such a consultative process given its credibility, both with developing country governments and with the global mining industry. In undertaking such an initiative, DTCD seeks to complement, and not duplicate, other international initiatives such as UNDP's Revolving Fund.

DTCD is of the view that there is a need for environmental *principles, guidelines* and *standards* regarding mining in developing countries. International principles would be of particular importance to development assistance agencies, while international guidelines would be of importance to the mining industry. Such guidelines could serve the function of clarifying allocation of responsibilities between the mining industry, the host government and multilateral and biliateral agencies. Such clarification could lead to more realistic expectations of the mining industry from environmentalists. Standards would form the basis for national environmental regulatory regimes.

DTCD's initiative is directed towards ultimately establishing general environmental guidelines relating to the different stages of the mining process, the different types of environmental impacts of mining and the health and safety impacts on workers and communities. The DTCD's initiative is also directed towards establishing specific environmental guidelines relating to the different types of mining, the mining of specific minerals, and the different technology processes involved in extraction, beneficiation and processing.

It is envisaged that the guidelines will be directed at different sets of actors, including developing country governments, mining companies and multilateral and bilateral development agencies. It is also envisaged that special attention will be devoted to identifying mechanisms for the implementation and monitoring of such guidelines.

9.2 DTCD's Inquiry: Mining Company Responses

DTCD raised seven issues with a large number of national and multinational mining companies. It received written responses from 13 companies (one each from Chile, England, Holland and Finland; two each from Japan, Canada and Australia; and three from the U.S.). The seven areas on which information was sought were:

- availability of company guidelines on environmental protection;
- importance of environmental protection in company strategy and planning;
- costs of environmental protection for the various types of mining and metallurgical operations;
- organizational set-up within the company for environmental protection;
- competitive impact of different environmental protection standards in various countries;
- special observations with respect to environmental regulations encountered in developing countries;
- importance of possibly less stringent environmental standards in developing countries in the choice of where to locate mining and metallurgical operations.

Of the 13 companies reporting, all but one had an environmental policy or 'philosophy', but only three (a Canadian, a Dutch and a Finnish company) had detailed environmental guidelines regarding mining. Of the three American companies reporting, two had almost identical environmental policies. This policy comprised four main commitments:

- to comply with all applicable laws and regulations regarding pollution control, land use, reclamation and environmental effects;
- to comply with all applicable public health and safety standards and to surpass them where technically and economically feasible to do so;
- to conduct research to develop new and more effective methods for environmental protection;

- to take reasonable measures to have company projects contribute positively to the economic and social well-being of their local communities.

The three European companies reporting also have broad environmental policies commiting (for example) to operate on best contemporary practice, to treat environmental protection as an integral part of production and to apply the best practicable means. The two Japanese companies claimed to have an environmental 'philosophy', and both saw commitment to the environment as being essential to the companies.

The two Australian companies rigidly subscribed to national environmental policies, and the National Conservation Strategy of Australia. Of the two Canadian companies, one is currently rewriting its policy on the environment, and is in the process of developing an environmental code of practice and defining roles and responsibilities. The other Canadian company has a policy that states that it "will strive to be exemplary leaders in environmental management by minimizing the environmental impact on the public, employees, customers and property, limited only by technological and economic viability".

Since only three of the companies reporting have environmental guidelines, each merits analysis.

The Canadian company's guidelines operate at two stages. At the Planning Stage the guidelines call for an environmental impact assessment, but only if required by the relevant governmental agency. The guideline also calls for an exploration and reclamation plan for sensitive areas such as permafrost areas, marshes and rivers. An overall reclamation plan is also required to be developed by the local manager for the entire target exploration area. At the Mining Stage, there are specific guidelines related to recreation activities, land clearing and buffer zone maintenance, road construction, campsite maintenance, sewage and waste management, fuel handling and storage, drilling operations and advanced exploration. Surprisingly, no guidelines are drawn up regarding the reclamation phase.

The Dutch company has prepared environmental management guidelines which apply generally to all companies within the group. The guidelines are intended for the guidance of these companies and they *may* consider using them in their operations. The guidelines identify four types of environmental impacts (physical, chemical, visual and odour) and categorizes them as causing: direct toxicological effects, indirect ecological effects, physical effects, or restriction of human uses. The guidelines create commitments to protect the environment from unacceptable damage and to minimize nuisance to the public, to adopt the best practicable environmental option, to comply with legislation setting minimum standards, to engage in constructive dialogue and to implement the environmental policies through effective management.

No specific targets are set but instead "measures of performance" are identified, such as the quantity of solid wastes produced and number of spills reported. Performance is to be judged during audits, appraisal visits, inspection or review meetings. The environmental management guidelines of this company lack sufficient specificity to deal with environmental problems associated with mining in developing countries. Its president stresses, "We must not promise what we cannot deliver. The group has grown not by shirking risk by by learning to manage it."

The North European company, consistent with its belief that "environmental protection is an integral part of production and belongs to the everyday work", has evolved environmental policies and practices (which it does not describe as "guidelines") to address specific environmental issues relating to mining. These practices include the creation of an environmental unit at each mine, delegation of powers and responsibilities to independent operating units, and the adoption of operational practices. These practices include daily calibration of all noise and vibration instrumentation, the setting of waste reduction targets, the adoption of pollution abatement techniques, internal education programmes, baseline studies and environmental impact assessment at the planning stage, research to develop problem-solving technologies, and improved monitoring mechanisms and procedures. This company's record provides a convincing example that practices evolved through sound environmental management can furnish an effective basis for the elaboration of environmental guidelines and standards.

All 13 of the mining companies which responded to the survey affirmed that environmental protection was a high priority in company strategy and planning. They reiterated the importance of incorporating environmental considerations in mine and plant design from the outset. One company noted that careful environmental planning can save money, lead to faster approvals, reduce liaison time and subsequent government monitoring. Careful attention to minimizing disturbance and subsequent rehabilitation can lead to more commercial opportunities.

The mining companies uniformly reported that cost information is not usually kept in a form to be extracted as requested. Various reasons (such as ore diversity, mine age, beneficiation and smelting, process differences, climatic conditions, proximity to urban centres) were given to explain why it is difficult to isolate or even estimate the environmental costs of mining operations. The wide variation in the figures presented make these difficulties obvious.

All of the reporting companies emphasized the importance of communications, and several described a structure which included corporate level environmental managers as well as site-specific environmental officers. Several companies stressed that environmental protection, like safety, is a line responsibility of each project officer. Policy-making, co-ordination, monitoring and audit functions were stressed, as was the importance of environmental education, training and retraining programmes. None of the companies indicated that financial constraints would impede expansion of staff to better implement environmental guidelines.

The reporting companies were unanimous in affirming that less stringent environmental standards in developing countries did not influence their investment or operational decisions. Factors identified as being much more important than the relatively small costs of environmental compliance were ore grade, mineability, accessibility, infrastructure, tax, currency considerations, royalty and licensing arrangements.

Similarly, several companies reported that differential environmental protection standards did not affect their business competitiveness or their business decisions because such companies, as a matter of corporate policy, always adopted their own higher national or corporate standards when they did business abroad. The Japanese and American respondents were, however, concerned about the competitive impacts of implementing environmental standards. Two points emerge from these divergent corporate views. First, none of the companies reporting is complaining about stringent environmental guidelines. Their only complaint is against national guidelines of differential stringency and uneven implementation thereof. The clear implication of the above is that companies favour the uniformity that international guidelines will contribute.

A number of the reporting mining companies made a call for international guidelines and standards to be applied in developing countries. There is also a need for an independent and knowledgeable agency to monitor performance based on these environmental guidelines. The general perception of international mining companies is that developing country environmental regulations vary from nil to very precise (but inappropriate) programmes. Moreover, even where regulations exist there are serious problems of ineffective or arbitrary enforcement.

From the above responses, it appears that international mining companies would support DTCD initiatives to strengthen national regulatory capacities in developing countries to enforce internationally agreed upon environmental guidelines on mining in those countries. One of the American companies stressed the need for greater input from the mining industry and from government agencies (such as, in the US, the Environmental Protection Agency, the Bureau of Mines, the Bureau of Land Management and the Bureau of Forest Services). It suggests that an industry review of the DTCD Draft Guidelines be co-ordinated by the American Mining Congress. According to this company, the guidelines should be flexible enough to allow for site-specific considerations.

9.3 DTCD's Inquiry: Reponse of Mining Industry Associations

DTCD has been furnished with copies of national industry-level guidelines from two countries. One (Canada) relates specifically to the mining industry. The other (Finland) relates generally to all Finnish industry, not just to mining.

The 'Guidelines for Environmental Care and Protection in Finnish Industry' is a concise general document, which was ratified by the Confederation of Finnish Industries in November 1988. The guidelines stress that "Minimizing the harmful effects on the environment is an indispensable part of industrial operations; cognisance of environmental issues is vital". In addition to the concept of 'product stewardship', the Finnish guidelines enunciate eight basic principles: holistic thinking (encompassing environmental aspects of the entire product cycle); recognising responsibility (of both the corporate organization and its personnel); duties of corporate management (including designation of resources); personnel know-how (developed through appropriate training); exceptional situations (to address emergency response and damage limitation); interaction with authorities (providing them with information); providing clients and customers with sufficient instructions; and active and informative communication with the company's own personnel, clients and society in general.

In June 1990, the Mining Association of Canada formulated its 'Guide to Environmental Practice' and pledged to "take every opportunity to further the objectives contained in the guide by communicating its content to member companies, to the mining industry at large, to other stockholders and to the public". The guide makes four commitments:

- to the concept of sustainable development which requires balancing good stewardship in the protection of human health and the natural environment with the need for economic growth;
- to high standards of management, both commercial and environmental, and to develop environmental skills for that purpose;
- to integrate economic and environmental decision-making in all activities and in all planning of future developments; and

- to find ways to measure and allocate the significant costs of long-term environmental protection.

The Canadian guide emphasises six principles:
Compliance. There should be full compliance with all of the applicable legisation providing for the protection of the environment, employees and the public.
Self-regulation. In the absence of legislation, companies will apply cost-effective, best management practices to advance environmental protection and to minimize environmental risks. However, the measures adopted must be technically and economically feasible.
Monitoring. Companies will maintain an active, continuing, self-monitoring programme to ensure compliance with government and company requirements. The emphasis is on self-monitoring rather than independent, external monitoring.
Research. Companies will make the result of non-proprietary research programmes available to the community at large in order to encourage the broadest possible application of research findings. However, the company itself is to be sole judge as to what research is non-proprietary.
Future Laws. Companies will work pro-actively with government and the public in the development of equitable, cost-effective and realistic laws for the protection of the environment. Moreover, the companies undertake to provide regulatory authorities with information to assess their environmental performance.
Communication. Companies will enhance communication and understanding with governments, employees and the public. While industry seeks opportunities to be heard, it recognises that others have the same right to objective and factual information.

The Canadian guide contains no specific standards or targets. It relies for its implementation solely on in-house monitoring programmes. The guide prescribes few specific obligations. So far as new mines are concerned, member companies will strive to use air, water and land prudently to conserve energy, to minimize discharges to air and water, to maintain the overall health of animals, fish and plant populations, to encourage recycling, to dispose of non-recyclable wastes in an environmentally sound manner, and to return disturbed land to a safe, stable and productive condition.

So far as existing mines are concerned, the Canadian guide recognises the obligation to take measures for improvement in existing mines, processing plants and waste management facilities but only where environmental benefits give value for money. Moreover, the Canadian guide states: "We accept our obligation to inform the public about mining successes and efforts to solve environmental problems". It is silent regarding failures.

So far as international mining industry efforts are concerned, two initiatives merit mention. The Mining Association of Canada has expressed support for 'The Business Charter for Sustainable Development: Principles for Environment Management' which was formulated by the International Chamber of Commerce in April 1991. That Charter adopts 16 principles; recognition of environmental management as a corporate priority, integrated management, process improvement, employee education, prior environmental impact assessment, environmentally safe and sensitive products and services, consumer advice, safe facilities and operations, research, precautionary approach, application of these principles to contractors and suppliers, emergency preparedness, transfer of technology, contributing to the common effort, openness to concerns, and compliance and reporting.

A second international initiative was launched in April 1991 when an international organization of mining companies was formed "to promote sound environmental and health policies and practices within the industry". Called the International Council on Metals and the Environmental (ICME), the group comprises industry leaders from Canada, France, Sweden, Australia, Germany, Mexico, Chile and the U.S. ICME's initial priorities include developing an environmental stewardship policy and working with the U.N. to create environmental guidelines for developing countries. ICME has created an advisory panel on communications, task forces to develop an environmental policy, and a standing committee on safe metals.

9.4 International Development Agency Guidelines on Mining

DTCD's draft guidelines seek to complement and not duplicate other existing international initiatives. Four such initiatives exist.

9.4.1 UNDP's environmental management guidelines

These guidelines are presently in their fourth edition and apply to UNDP country programmes and all UNDP financed activity. Part I of UNDP's 'Principles and Guidelines for Environmental Assistance and Sustainable Development in Technical Assistance' presents the basic concepts and ideas relevant to environmental management. Part II provides operational guidelines directed at UNDP personnel and Part III provides guidelines for environmental assessment and management regarding specific aspects of the environment.

The UNDP guide emphasizes institution-building, community participation and roles of non-governmental organizations. In the manufacturing and service sector, the guide adopts a 'cradle to grave' approach in which all steps of manufacturing a product are taken into consideration, rather than only focusing on controlling polluted effluents at the last step of manufacturing. Four environmental management tools are identified; environmental overviews of country programmes, environmental overviews of country projects, environmental screening criteria, and specific environmental management strategies.

The UNDP guide devotes attention to alternative means of enforcement of environmental policies and regulations (e.g. ethical persuasion, direct legal regulation and economic incentives) and introduces the interesting concept of environmental mediation.

9.4.2 UNDP – UNRFNRE guidelines

Of direct relevance is the UNDP initiative to create a UN Revolving Fund for Natural Resources Exploration (hereafter the Fund). Environmental protection guidelines have been drawn up by the

Fund covering health and safety, environmental impacts, and socio-cultural impacts. The guidelines cover both the exploration stage as well as all aspects of the mining process.

Pollution of surface and groundwater, acid mine drainage, radiation and methane receive special treatment, as do several other environmental impacts relating to air pollution, aesthetic pollution, reclamation and restoration. For all projects assisted by the Fund, an environmental overview (at the earliest stage), environmental impact assessment (to be jointly undertaken by the host government) and an environmental management plan must be drawn up, both at the exploration stage and at the stage of feasibility studies for mining projects. Post-project environmental evalution is also required.

9.4.3 UNEP technical guide

This guide deals with the four most common non-ferrous metal ores (copper, nickel, lead, zinc) as well as with gold. The guide asserts two basic propositions. First, that technology and control procedures do exist to reduce the environmental impacts of mining to acceptable levels. Second, that issues related to the working environment are closely linked to the general environment. The guide proposes that sanctions be taken against management who do not adhere to the regulations concerning the provision of a safe working environment.

The guide recommends that a specific international mineral industry safety code be developed. Typical problems associated with mining such as acid mine drainage, air pollution, noise, vibrations, tailings and waste rock disposal and reclamation merit special treatment in the guide. Specific measures recommended by the guide include environmental assessment, environmental management, pollution control strategies, pollution monitoring (including sampling), training for safety and accident prevention, health and safety operational guidelines, and emergency response plans.

9.4.4 World Bank — draft policy paper

The World Bank is presently preparing a draft policy paper (hereafter Bank paper) on 'Mining and Mineral Processes: Potential Negative Impacts and Mitigating Measures'. The Bank paper contains a careful detailing of the potential environmental impacts of several different types of mining activities; surface, underground, dredging and placer mining, heap leaching, *in situ* and *ex situ* processing. The Bank paper identifies specific natural resource issues involved, the socio-cultural issues involved (e.g. changes in land use) and special issues relating to reclamation.

The Bank paper emphasizes the importance of management, monitoring and training. But perhaps the most significant contribution of the Bank paper lies in its careful identification of a range of mitigative measures that could be taken. "Other than the alternative of 'no action', or not going forward with all or part of the project, alternatives for the mining (project) are generally in the type and degree of mitigation. Mitigation may be specially tailored to the particular operation."

All of the above international initiatives are of undoubted value and significance. But they are operational only in cases where the respective development assistance agency (UNDP, World Bank) is directly involved in the project. Like the UNEP guidelines, the DTCD initiative seeks international approval and national implementation of its draft guidelines in all projects involving minerals' development whether or not bilateral or multilateral development assistance is involved.

9.5 Towards International Standards

The process of securing international adoption of principles and guidelines is a difficult and protracted one in any field. It is likely to be especially so in respect of environmental aspects of mining in developing countries. Moreover, such a process will, of necessity, require participation by a diverse range of actors including host country governments, mining companies, development assistance agencies, trade unions, and environmental organizations.

There is room for optimism, however, because of the recognition, both by the mining industry and by the development assistance community (multilateral and bilateral) that the national and global environmental impacts of mining and minerals' development programmes in developing countries merit the highest priority. The DTCD inquiry provides ample documentation of such recognition. Hence, the process of adopting international principles and guidelines on environmental aspects of mining in developing countries might well prove less arduous than expected. But for this to be so, it is very important that other items on the agenda for global negotiations should not be brought into what are essential negotiations over environmental responses to environmental concerns. Important, but unrelated, issues should not complicate the process of adoption of environmental guidelines regarding mining in developing countries.

Six key issues, relevant to such negotiation process, need to be highlighted and addressed:

Confidentiality and The Right to Know. There is repeated reaffirmation that environmental considerations must enter at the earliest possible stage of minerals' development activities. But there are genuine needs for confidentiality that need to be balanced against needs for public disclosure.

Costs of Environmental Compliance. Despite mining company tact on the subject, there is no doubt that environmental costs are real, and need to be met collectively by mining companies, host governments, development assistance agencies and consumers. Mechanisms need to be developed to meet these costs; both long-term pollution and short-term remedial environmental costs.

Impact Assessment and Management Programme. The monitoring and implementation of environmental guidelines is almost as important as the articulation of those guidelines. There are very important problems relating to the above two management tools which all interest groups agree are vital to effective environmental regulation. Serious practical problems exist about the successful conduct of impact assessment and management programmes — who does the job? how are the costs met? how to ensure credibility of the process and its outcome? how to ensure participation? and accuracy?

Reconciling Conflicting National Priorities. Environmental priorities (especially long-term ones) may often assume global but not national significance in developing countries faced with immedate-term concerns regarding exchange and revenue earning and employment. Development assistance agencies have vital roles to play in reconciling and adjusting such conflict in priorities.

Reconciling Conflicting Interests. Mining in developing countries involves an inherent and basic conflict between governments (who wish to maximize economic return) and foreign mining companies (who wish to optimize profits). Environmental considerations must not be held hostage to such conflicts. Environmental mediation mechanisms and multilateral and bilateral development assistance agencies both have crucial roles to play in this regard.

Participation by Concerned and Affected Groups. Many, if not most, of the problems arising from health, safety and environmental impacts on affected communities could be averted if significant, effective and early participatory mechanisms could be evolved. Often, a foreign mining company becomes a scapegoat for a host government which has opted for bypassing meaningful local community participation procedures. This is a classic 'no win' situation which needs to be addressed by policies which do not fear or begrudge participatory processes.

CHAPTER TEN

The Berlin Guidelines

As the Berlin Round Table drew to a close, a series of discussions were held to obtain a consensus of the views amongst participants at the meeting. This chapter comprises the full document submitted.

Worldwide long-term economic development can best be achieved through the pursuit of sustainable development policies comprising a balance of economic, socio-cultural and environmental protection measures. While taking into account global environmental concerns, each country should apply this concept to meet the needs of its environment and economic circumstances.

Sustainable mining activities require good environmental stewardship in all activities, from exploration and processing to decommissioning and reclamation. It acknowledges the importance of integrating environmental and economic considerations in the decision-making process and the fact that the mineral deposits are unique in their occurrence. It recognizes the importance of mining to the social, economic and material needs of society, in particular for developing countries, and that minerals, notably metals, offer great potential for the use of future generations through increased recycling programmes.

Sustainable mining under appropriate environmental guidelines is based on interaction between industry, governments, non-governmental organizations and the public, directed towards optimizing economic development while minimizing environmental degradation. The need for such guidelines is recognized by industry, governments and international agencies. It is also recognized that the political will of governments, together with the commitment of industry management and of the community, are the essential conditions needed to enforce environmental legislation and, more importantly, to ensure compliance with all applicable laws for the protection of the environment, employees and the public.

10.1 Addressed to the Mineral Sector

Governments, mining companies and the minerals industries should as a minimum:

1. Recognize environmental management as a high priority, notably during the licensing process and through the development and implementation of environmental management systems. These should include early and comprehensive environmental impact assessments, pollution control and other preventive and mitigative measures, monitoring and auditing activities and emergency response procedures.
2. Establish environmental accountability in industry and government at the highest management and policy-making levels.
3. Encourage employees at all levels to recognize their responsibility for environmental management and ensure that adequate resources, staff and requisite training are available to implement environmental plans.
4. Ensure the participation and dialogue with the affected community and other directly interested parties on the environmental aspects of all phases of mining activities.
5. Adopt best practices to minimize environmental degradation, notably in the absence of specific environmental regulations.
6. Adopt environmentally sound technologies in all phases of mining activities and increase the emphasis on the transfer of appropriate technologies which mitigate environmental impacts, including those from small-scale mining operations.
7. Seek to provide additional funds and innovative financial arrangements to improve environmental performance of existing mining operations.

8. Adopt risk analysis and risk management in the development of regulation and in the design, operation and decommissioning of mining activities, including the handling and disposal of hazardous mining and other wastes.

9. Reinforce the infrastructure, information systems service, training and skills in environmental management in relation to mining activities.

10. Avoid the use of such environmental regulations that act as unnecessary barriers to trade and investment.

11. Recognize the linkages between ecology, socio-cultural conditions and human health and safety, both within the workplace and the natural environment.

12. Evaluate and adopt, wherever appropriate, economic and administrative instruments such as tax incentive policies to encourage the reduction of pollutant emissions and the introduction of innovative technology.

13. Explore the feasibility of reciprocal agreements to reduce transboundary pollution.

14. Encourage long-term mining investment by having clear environmental standards with stable and predictable environmental criteria and procedures.

10.2 Addressed to Development Assistance Agencies

Multilateral and bilateral assistance agencies have an essential role to play in furthering environmental management, particularly in developing nations, and in assisting these nations in programmes to protect their environment, both nationally and as part of the global environmental system. Accordingly, they should:

1. Accord high priority to the mitigation of environmental degradation associated with mining in developing countries to achieve high environmental performance.

2. Initiate, as an integral part of any exploration and mining project, environmental institutional building programmes. Special support should be given to countries actively working to improve their environmental capabilities.

3. Require that all mining projects supported shall contain a training component that will include specific training on environmental awareness and its application to the mining sector.

4. Support increased research regarding the development of new processes, with fewer environmental impacts, including recycling.

5. Support the development of activities that would mitigate adverse effects on the socio-cultural fabric and the ecosystem. To achieve this objective, international agencies should give priority to education and training which increase awareness of these issues and allow the affected communities to participate in decision-making.

6. In supporting mining projects, agencies should also take into account the following:

- rehabilitation of displaced population;
- environmental history of the country;
- large scale impact on socio-cultural patterns;
- the overall economic balance of the project vis-à-vis its total environmental impact;
- the impact on other natural resources and ecologically sensitive areas (e.g. protected forest lands, mangroves, wildlife parks and neighbouring waterbodies).

7. Promote conferences and policy research on environmental management practices and technologies, and ensure the dissemination of this information.

8. Support and promote regional co-operative programmes to achieve sustainable development of mineral resources.

9. Adopt environmentally safe methods of mining and processing for existing projects.

10. Increase and co-ordinate their assistance to developing nations in the field of environmental policies management.

APPENDIX A

SUMMARY OF DISCUSSIONS AND PAPERS

The Berlin Round Table was opened by the chairman of the meeting, Heino E. Wittrin, who introduced Rudolf Vieregge, director of Environmental Planning and Policies, who was standing in for an unavailable Clemens Stroetmann, the State Secretary in the Federal Ministry for the Environment, Nature Conservation and Nuclear Safety. Mr Vieregge talked of the effects of environmental problems on international trade, and the export of pollution problems. Reliance on the General Agreement on Trade and Tariffs (GATT) was not enough, and Mr Vieregge spoke of the danger of unfettered world trade and conflicts of interest. He stressed that the world needs 'predictable' regulations and multilateral standards.

Dunja Pastizzi-Ferencic also welcomed the delegates in her role as director of UN/DTCD's Natural Resources and Energy Division. Ms Pastizzi-Ferencic spoke of the Stockholm Conference in 1982, and the establishment of a mandate for environmental awareness. She noted that the main purpose of the Berlin meeting was to formulate international principles, guidelines and national regulatory standards regarding mining. There needed to be a particular emphasis on the initiation of a consultative process between the various 'actors' involved. Ms Pastizzi-Ferencic called for better co-ordination and co-operation among, and between, bilateral and multilateral development agencies in respect of mining and the environment activities in developing countries.

Ms Pastizzi-Ferencic reminded delegates that environmental issues have continued to evolve since the World Commission on Environment and Development (also known as the Brundtland Commission) called for decisive political action to link economic growth with sound management of the environment. The UN offered a crucial role in translating the rhetoric of sustainable development into operational standards against which state policies can be evaluated.

Ms Pastizzi-Ferencic and Beatrice Labonne, chief of the UN/DTCD's Mineral Resources Branch, argued that there was a general consensus that three sets of linkages needed to be addressed to establish a bridge between the environment and development. These were related to poverty and demographic pressures; economic growth patterns, consumption standards and the environmental consequences; and international debt adjustment and deterioration in terms of trade.

The working sessions were begun by Chris Hinde of London-based Mining Journal Ltd, who gave an overview of the role of mining, including a brief mention of its importance to developing countries in terms of both trade and infrastructure. He also stressed the industry's role in the lubrication of international trade. In an overview of some of the environmental issues he covered the different impacts on the environment of old and new mines or smelters.

Dr Hinde also noted that he believed the instigators of the problems, on a tonne-for-tonne basis, were the smaller operations and artisan miners. After briefly introducing the eleven case studies (presented in a report by Mining Journal Research Services) he concluded by saying that the industry needed increased communication. This would include a comprehensive database, incorporating the costs and techniques of existing applications. Dr Hinde also noted that the mining industry needed to launch a co-ordinated bid to capture the initiative — to set, rather than to constantly respond to, the agenda for the burgeoning debate on the environment.

Ricardo Antunes, a manager from Companhia Vale do Rio Doce (CVRD) in Brazil, stressed that garimpo workings in his country were not simply a flight from poverty, and that more education was needed. In giving examples of small-scale mining in Ghana, Kwabena Mate, a senior legal officer from Ghana's Minerals Commission, explained that workings were divided into supervised districts, and that all of the gold produced was sold through the government, and a 3% deduction was charged for future reclamation.

Jaime Solari, co-ordinator with Santiago's Ministry of Minerals, explained that the situation in Chile was different from both Brazil and Ghana. There was little pollution from 'gold-rush' mining. Rather it was the state-owned companies which had been the worst culprits. He stressed that an equilibrium was needed between economic growth and the environment. Different time scales would be required according to individual countries' circumstances. Another delegate from Latin America, Guillermo Cortez, an adviser with the Ministry of Minerals and Metallurgy, described the situation in Bolivia as "not easy" because of the closure of numerous large mines and the difficulty in controlling smaller operations. Mr Cortez spoke of the need for guidelines only, because each country had a different scenario. Martin Pretzell, counsellor with the German Federal Ministry for Economic Co-operation, agreed, adding that legislation was not enough. Education was necessary to create the required awareness of the environment.

In her paper, which opened the Second Session of the Round Table meeting, Alyson Warhurst noted that regulation assumed static technology and was difficult to enforce. Consequently, regulation had not been successful when measured in terms of reductions in pollution. Indeed, Dr Warhurst argued that regulation has had a negative effect on advances in technology. Her paper suggested that economic growth and environmental sustainability of mineral production in developing countries will require improved operating efficiencies (in turn dependent upon the extent of technology transfer and training). This development can be aided through designing country-specific flexible regulations.

In congratulating Dr Warhurst on her paper, Thomas Walde questioned whether the developing countries were able to reach a satisfactory level of environmental protection without the help of international mining companies. Several delegates suggested that developing countries should build an increased element of technology transfer and training into any agreements with foreign companies.

The perspective from developing countries on environmental policies was the subject of a paper from Bharat Dhar of the Benares Hindu University in India. Professor Dhar outlined the uniqueness of mining and the need for guidelines within a broad framework. He stressed that the policies imposed by the lending agencies must be in line with the country's own aims. The merit of evolution of regulation from within each country, rather than external decree, was discussed. The transfer of technology was also an important subject of debate. Examples of equipment which were too sophisticated for the local level of expertise were given. This problem was expanded on by several participants, who stressed the need for developing countries to acquire technical capacity. This can only come through training to adapt and innovate, rather than simply being taught to use a new piece of equipment.

The second full day of the Round Table began with a presentation on the economic aspects of environment protection by Beatrice Labonne. She focused upon the comparative costs of environmental protection, the impact of costs on international competitiveness, and the economic and financial considerations influencing the environmental aspects of mining.

Richard Westin, from the University of Houston Law Center, presented a paper on the tax treatment of environmental protection measures. Mr Westin discussed microeconomic theories of taxation, and stressed the need to co-ordinate environmental and tax policy-making. Debate focused upon the effect of tax charges, and whether deductions should be made to cover the cost of ultimate renovation.

In a joint paper, Thomas Walde and Lawrence McDonnell (from the Universities of Dundee and Colorado, respectively) discussed legislation and regulations concerning environmental protection. The necessity of an integrated approach to all the various policy instruments was noted. In particular the authors argued that it was important that social interests were safeguarded and that regulations were combined with market forces to achieve the objectives.

Allen Clark of the East-West Centre in Hawaii presented a paper on the socio-cultural impact of mining in developing countries. He argued that, because of the enclave nature of most mining ventures in developing countries, any new mining operation will cause some degree of disruption to society. A mine will spawn a series of service industries in the area, providing opportunities for local entrepreneurs. The native people need to be properly prepared for these changes.

Mr Clark also noted that any new mine may create cultural tension and considerable strain on traditional support systems. Local inhabitants, lacking the necessary skills, may resent their exclusion from key jobs. Careful planning is also needed to maintain some means of social support when the mine is decommissioned. Guidelines were suggested to minimize the negative socio-cultural impact of mining on rural populations.

Roderick Eggert, associate professor at the Colorado School of Mines, gave a paper on the 'optimal amount' of protection for the environment. Mr Eggert emphasized that environmental protection comes at a cost, and that this cost must be commensurate with the expected benefits. Optimally, realistic goals should be pursued rather than striving to eliminate pollution altogether. The different roles of national governments and international agencies were discussed.

The final paper of the Round Table was presented by Ayesha Dias, of the UN/DTCD, with a review of the role of international co-operation agencies (both multi- and bi-lateral) in helping developing countries to meet the environmental challenge. Delegates discussed the various means by which the effectiveness and impact of technical assistance projects could be improved.

SUMMARY

The issue of the sustainability of the mining sector was an important aspect of much of the debate at the Round Table. Several participants described the total cost of mining as including those charges associated with environmental control and restoration. However, perceptions from the developing countries differed from this economic concept. Other issues were outlined, including the socio-cultural factors, employment and the need to generate foreign exchange. These factors impinge upon the size of the operations and the tolerance of artisan workings.

It was noted that environmental protection in mining needs to be site specific, and that environmental regulations are mostly applied to large-scale operations. Important topics were the size of mining operations and the relative merits of regulations, financial incentives and the enforcement of environmental guidelines. The issues raised can be conveniently grouped under three general headings: local, national and international.

1. LOCAL ISSUES

Employment: Garimpo (gold-rush or artisan) workings are important local providers of employment in developing countries. Similarly, despite the problems of control, many small-scale operations are important to local economies and cannot be easily shut down. Meanwhile, increased mechanization (often encouraged by environmental considerations) on the larger mines is leading to lower levels of employment. This may encourage workers into illegal operations and/or into the small-scale mining sector. There is a need to channel properly the income of small-scale mining into the national economy.

Control: Larger mining operations are more easily controlled and monitored, and they can more adequately cope financially with environmental regulations. It was noted that very few international companies would fail to follow guidelines if these were endorsed by the international community. Ms Labonne was one of several participants who believed that it was artisan workings rather than small-scale mining which were of concern. Several speakers believed that small-scale mining could not be improved by stricter legislation, and that governments needed to help in the formation of bigger units through the creation of co-operatives and central processing. Several participants also believed that there is merit in making artisanal operations legal so that they can be brought into a control framework. It is also important to demonstrate to the smaller mining companies the advantages of implementing technological measures to improve recovery rates and environmental performance. In most countries the difficulty is the inefficient use of existing technology and the problem of enforcing regulations. However, there was no consensus at the Berlin meeting on the imposition of minimum pollution levels for problems of a purely national nature.

Costs: The Round Table noted that the cost of implementing environmental measures for small-scale mines was proportionally higher to revenue than that for the larger operations. It was unrealistic to expect marginal operations to accept this burden; it would only result in a disregard for the law, or closure of the operation. Neither do many of these smaller operations have access to the necessary level of funding.

Trust Funds: The concept of providing well-secured funds towards the costs of future environmental restoration was noted. Examples were given of the successful use of Trust Funds for rehabilitation, including Australian uranium mines (whose funds are held by the government), and U.S. nuclear decommissioning funds and locally-orientated U.S. coal trusts. A 'sinking fund' approach to calculate the contributions was suggested.

Socio-Cultural: The impact of mining on the socio (between groups) and cultural (group values) aspects of local communities was emphasized. The disruptions occur throughout the life cycle of the mining enterprise but are most intense in the early phases — resulting in conflict within families and communities, and are largely irreversible. The problems were identified as having local, regional and national impacts. These socio-cultural aspects are neither well-recorded nor monitored.

2. NATIONAL ISSUES

Technology Transfer: The conference discussed the transfer of technology, and the importance of comprehensive training and information exchange being included in the package. This is because skills and technical knowledge are required to adapt, maintain and improve plant and equipment to ensure continued production efficiency and sound environmental management.

Training & Education: This was the subject of much debate. According to participants, there remains a serious shortage of the necessary environmental management skills in many of the developing countries. Training programmes need to be implemented with the target of establishing an aptitude for innovation. The process of education would be slow but should become part of all future joint venture agreements with multinational companies. There was a need to design training programmes aimed at enhancing technology management.

Legislation: Numerous speakers pointed to the failure of legislation because of the difficulty of enforcement, particularly with regard to smaller operations. The benefits of an increase in the size of operations through co-operatives and zoning, with government organized central administration, was mooted. Existing problems with 'command and control' systems were discussed. It was doubted whether regulations, on their own, would result in an optimal reduction of pollution. Legislation should be reached through concensus, and be part of a package of instruments which would include financial incentives and a timetable for the adoption of mitigating measures.

Economic Instruments: Fiscal advantages were believed to be essential in any workable scheme to improve the environment. The establishment of rehabilitation and personal health funds was widely recommended. These should enjoy tax benefits and needed to be kept secure. Direct grants were mentioned as an option for smaller producers. The trade of emission permits was also discussed as a tool to achieve the overall reduction of pollutant emissions. In all cases the fiscal regime should be flexible and remain relevant.

3. INTERNATIONAL ISSUES

International Co-operation: There was a call for better communication between mining companies and international agencies. This would make better use of existing expertise and might promote a better understanding of mining and environmental issues. The importance of interaction and co-ordination between governments, companies and the local community was also emphasized.

International Agencies: It was noted that organizations such as the UN/DTCD had an important role as providers of information (on environmental costs and technology in particular), and for assistance in drafting national guidelines. (In this respect, simpler communication channels would be welcomed.) The international agencies were also thought to have a role to play in the restoration of abandoned mining sites. However, it was noted that no economically viable project, provided it was located in a country with a conducive investment framework, had failed due to a lack of funds — it was the marginal projects which suffered.

Short-Term Finance: The role of international agencies was stressed with regard to financial support. There were also examples of assistance at the national level; including Morocco's provision of up to 50% of the cost of any environmental work.

Polluter Pays: There was debate on the justice of developing countries facing onerous environmental restrictions. The point was made that the industrialized countries, which are still the largest source of pollution, should not inflict excessive environmental costs on less economically advanced nations. It was noted that, in the future, courts (in any country where a company has assets) may be empowered to award damages against pollution (of a global nature) in another country.

Trade Barriers: The increase of environmental barriers to trade was discussed. Where pollution was of an international nature, one speaker anticipated that import restrictions would occur.

Environmental Audit: Participants discussed both the advantages and disadvantages of establishing independent consultancies able to advise on environmental matters, conduct environmental audits and monitor mine performance.

Guidelines: The development of internationally acceptable guidelines was discussed (based on a report prepared by UN/DTCD entitled 'Towards Environmental Guidelines on Mining in Developing Countries'; see Chapter 9). A first document was produced at the Berlin meeting, and this is reproduced as Chapter 10 of this book.

APPENDIX B

LIST OF PARTICIPANTS

1. International Participants

ANGOLA
Augusto Germano de Araujo
Director
Servico Geológico de Angola
Instituto Nacional de Geologia
C.P. 1260
Luanda

BRAZIL
Ricardo Antunes
Manager
Companhia Vale do Rio Doce
Av. Graça Aranha, 26/12° andar
Rio de Janeiro — RJ 20005

Observer:
J. Osael Farías
Manager
Rio Doce International S.A.
Av. Louise 326, Btes 29/30
B 1050 Bruxelles

BOLIVIA
Guillermo Cortez A.
Adviser
Ministerio de Mineria y Metalurgia
Av. 16 de Julio
Casilla 8686
La Paz

CHILE
Jaime Solari
Co-ordinador
Unidad Ambiental
Ministerio de Mineria
Teatinos 120, Piso 9
Santiago

CHINA
Zhang Hongren
Chief Engineer
Ministry of Geology and Mineral Resources
64 Funei Street
Beijing

COLOMBIA
Ibrahim Abu-Abed
GTZ Consultant to Corponariño
Proyecto de Cooperación Técnica
Colombo-Alemana (Coteca)
Apartado Aéreo 902
Pasto

GHANA
Kwabena Mate
Senior Legal Officer
Minerals Commission
The State House
P.O. Box M 248
Accra

HUNGARY
Arpád Bajkay
Head of Section
Environmental Management
Ministry of Industry and Trade
Iparies Kereskedelmi Ministerium
Mártirok utja 85
H 1024 Budapest II

INDONESIA
Kosim Gandataruna
Director General of Mines
Ministry of Mines and Energy
Jl. Gatot Subroto 49
P.O. Box 390/Kby
Jakarta, 12950

MOROCCO
Mohamed Karbid
Secrétaire Général
Ministère de l'Energie et des Mines
Haut Agdal
Rabat

NAMIBIA
Sybrand A. de Waal
Deputy Permanent Secretary
Ministry of Mines and Energy
P.B. 13297
Windhoek

NIGERIA
Jacob O. Ameyan
Assistant General Manager
Federal Environmental Protection Agency
Federal Secretariat
Lagos

PAKISTAN
Khawaja Asifullah
Director
Pakistan Mineral Development Corporation
13, H-9
P.O. Shaigan
Islamabad

PAPUA NEW GUINEA
Nilip P. Nakikus
Counsellor
Embassy of Papua New Guinea
Gotenstrasse 163
D 5300 Bonn 2

PERU
Oscar Bernuy Verand
Director General de Mineria
Ministerio de Energia y Minas
Av. las Artes No. 260, San Borja
Lima

PHILIPPINES
Victor O. Ramos
Under-Secretary
Department of Environment and Natural Resources
Visayas Avenue
Diliman, Quezon City

THAILAND
Thasanee Chantadisai
Chief of the Agriculture and Mining Section
Environmental Impact Division
National Environment Board
60/1 Soi Phibun Watthana 7
Rama VI Rd.
Bangkok 10400

2. German Participants

Friedrich von Bismarck
Senior Consultant (Mining)
Maienstrasse 4
D 1000 Berlin 30

Ortwin Caldonazzi
Head of Section
Vereinigte Mitteldeutsche
Braunkohlenwerke-AG
Postfach 61
O-4400 Bitterfeld

Werner Efing
Member of the Managing Board
Deutscher Braunkohlen-Industrie-
Verein
Max Planck-Strasse 37
D 5000 Köln 41

Klaus Germann
Professor, Technical University Berlin
Sekr. BH 4
Ernst Reuter-Platz 1
D 1000 Berlin 12

Wolfgang von Hansemann
Mining Engineer
Deutsche Gesellschaft für
Technische Zusammenarbeit
Postfach 5180
D 6236 Eschborn 1

Thomas Hentschel
Geologist
Projekt-Consult GmbH
Limburger Strasse 28
D 6240 Königstein

Christina Kammlage
Carl Duisberg Gesellschaft e.V.
Lützow-Ufer 6
D 1000 Berlin 30

Christoph Kippenberger
Bergdirektor
Bundesanstalt für Geowissenschaften
und Rohstoffe
Postfach 510153
D 3000 Hannover 51

Michael Priester
Consultant
Beratung in Entwicklungsländern
GmbH
Limburger Straße 28
D 6240 Königstein/Ts.

Christof Rotberg
Director of Section
Deutsche Investitions
und Entwicklungsgesellschaft mbH
Belvederestraße 40
D 5000 Köln 41

Michael Schmidt-Thomé
Head of Section
Bundesanstalt für Geowissenschaften
und Rohstoffe
Postfach 510153
D 3000 Hannover 51

Wolfgang Siegel
Director
Kreditanstalt für Wiederaufbau
Palmengartenstraße 5-9
D 6000 Frankfurt am Main

Michael Strzodka
Head of Section
Lausitzer Braunkohle AG
Laugktrasse 15
O-7840 Senftenberg

Wolfgang Wendt
Missionswerk der
Evang.-Luth. Kirche in Bayern
Postfach 68
D 8806 Neuendettelsau

3. Representatives of International Organizations, Associations and Industry

Gary Nash
Secretary-General
International Council on Metals
and the Environment
360 Albert Street, Suite 1550
Ottawa, Ontario K1R 7X7
Canada

Lutz A. Baehr
Director
Management Team
United Nations Centre for Science
and Technology for Development
Room 1078
One, United Nations Plaza
New York, N.Y. 10017
U.S.

Norman Jennings
Mining Industry Specialist
International Labour Organisation
Industrial Activities Branch
4 route des Morillons
CH 1211 Genève 22
Switzerland

Lorraine Ruffing
Chief, Accounting and Statistics
United Nations Centre
on Transnational Corporations
One United Nations Plaza
DC 2, Room 1220
New York, N.Y. 10017
U.S.

Hennie Veldhuizen
Vice-President
Noranda Minerals Inc.
Environment
4 King Street West
Toronto, Ontario M5H 3X2
Canada

Olle Östensson
Senior Economist Commodities
Division
United Nations Conference on
Trade and Development
Palais des Nations
CH 1211 Genève 10
Switzerland

Fritz Balkau
Senior Progr. Officer
United Nations Environment Programme
Industry and Environment Office
39-43 quai André Citroën
F 75739 Paris Cedex 15
France

Ernst Bolte
Senior Economist Metal and Mining Unit
The World Bank (IBRD)
Africa Technical Department
1818 H Street, Room 8061
Washington, D.C. 20433
U.S.

Bernt Rosholt
Operations Supervisor Geologist
United Nations Revolving Fund for Natural Resources Exploration
Room 9116 FF, One UN Plaza
New York, N.Y. 10017
U.S.

Richard Tallboys
Chief Executive
World Coal Institute
Vicarage House
58-60 Kensington Church Street
London W8 4DB
U.K.

4. Panel Members

Ernst Ulrich von Weizsäcker
Director
Institute for European Environmental Policy
Aloys-Schulte-Straße 6
D 5300 Bonn

Claus N. A. Siebel
Chairman
German Association of International Mining
Zitelmannstrasse 9-11
D 5300 Bonn 1

Keith C. Hendrick
Chairman
Noranda Minerals, Inc.
Suite 900
4 King Street West
Toronto
Canada M5H 3X299

Hugh M. Morgan
Managing Director
Western Mining Corporation Ltd
360 Collins Street
Melbourne
Australia

5. Resource Persons

Allen L. Clark
Programme Director
Minerals Policy Program
East-West Centre
12124 Quorn Lane
Honolulu, Hawaii 96848
U.S.

Bharat Dhar
Department of Mining Engineering
Institute of Technology
Benares Hindu University
Varanasi 221005
India

Roderick G. Eggert
Associate Professor, Mineral Economics Department
Colorado School of Mines
1500 Illinois Street
Golden, Colorado 80401
U.S.

Lawrence J. MacDonnell
Director, Natural Resources Law Center
University of Colorado School of Law
Campus Box 401
Boulder, Colorado 80309-0401
U.S.

Thomas Wälde
Executive Director
Centre for Petroleum and Mineral Law Studies
University of Dundee
Dundee DD1 4HN, Scotland
U.K.

Alyson Claire Warhurst
Fellow
Science Policy Research Unit
OECD Mining and Environmental Policy Project
Brighton BN1 9RF, East Sussex
U.K.

Richard Axel Westin
University of Houston Law Center
4800 Calhoun
Houston, Texas 77004
U.S.

6. Representatives from the Government of the Federal Republic of Germany

Rudolf Vieregge
Director, Department of Environment Policy and Planning
Federal Ministry for the Environment, Nature Conservation and Nuclear Safety (BMU)
Kennedyallee 5
Co-ordination
D 5300 Bonn 1

Martin Pretzell and Gerhard Lüdtke
Counsellors
Federal Ministry for Economic Co-operation (BMZ)
Karl Marx-Strasse 4-6
D 5300 Bonn 1

7. Organisers

German Foundation for International Development (DSE)
Villa Borsig
Reiherwerder
D1000
Berlin 27, Germany

Heino Wittrin
Chairman of the meeting
U.N. Associate Director General (retired)

Götz Link
Executive Director DSE

U.N. Natural Resources and Energy Division
Department of Technical Co-operation for Development
One United Nations Plaza
New York, N.Y. 10017-3514
U.S.

Dunja Pastizzi-Ferencic
Director

Beatrice Labonne
Chief, Mineral Resources Branch

Daniel Vock
Interregional Adviser on Mineral Legislation and Investment Promotion

Ayesha Días
Expert

Antonia Cubeiro
Economic Affairs Officer
Mineral Resources Branch

8. Other Representatives

Mining Journal Ltd
60 Worship Street
London EC2A 2HD
U.K.

Christopher Hinde
Editorial Director

Nathalie Rosin
Assistant Editor

APPENDIX C

ANNOTATED AGENDA

Tuesday, June 25, 1991

Working Session I

Comparative survey of environmental problems facing mining and metallurgical operations
(Introduction by Dr Chris Hinde, Mining Journal)

The survey describes the inter-relation between the minerals industry and the environment, and highlights some of the measures that have been taken by the various parties involved. Some of the studies look at cases where clearly a problem exists, while others report on operations where measures have been taken to ensure that the environmental impact is minimised. The concerns of governments are shown through the introduction of new regulatory procedures. Case studies are provided for:

1. Carajas Iron Ore Project, Brazil.
2. Pueblo Viejo, Dominican Republic.
3. Brown Coal Industry, Eastern Germany.
4. Government Policy, Ghana.
5. Black Angel Lead-Zinc Mine, Greenland.
6. Bauxite and Alumina, Jamaica.
7. Revegetation of Tin Mine Tailings, Malaysia.
8. Rössing Uranium Mine, Namibia.
9. Mercury Hazards, Philippines and Brazil.
10. National Environmental Strategy, Poland.
11. Kola Nickel, Soviet Union.

Working Session II

Evolution of Environmental Policies for the Mining Industries in Developing Countries
(Introduction by Dr Alyson Warhurst, OECD Mining and Environmental Policy Project, Sussex)

Environmental degradation is not a necessary consequence of mining. Rather, it is the socio-economic organisation of mining production that influences environmental behaviour, in addition to the role that local geology, geography and climate play in determining the extent of the hazard created. Regulatory frameworks, though growing in complexity, will not alone solve environmental degradation problems, particularly not in the developing countries, principally because production efficiency is a precursor to achieving environmental efficiency. Moreover, enforcing regulation requires skills and resources beyond the budget and capabilities of many developing countries. Far-sighted companies are innovating and changing the broader parameters of mineral production. Contrary to the belief that a trade-off exists between production costs and enviromental costs, which presupposes a static technology, they are building into new generations of technology imported economic and environmental costs.

Dr Warhurst suggested that participants discuss:

1. How do existing 'command and control'-type regulations tend to deal with the symptoms of environmental mismanagement (i.e. pollution) rather than its causes (poverty, lack of access to technology and capital, human resource constraints, etc.)?
2. What will be the cost of prospective environmental security?
3. Could such investments herald another wave of technological dependence – the swapping of one social cost for another to the benefit of foreign capital (reinforced by credit conditionality), or will smelting and leaching alternatives be developed, adapted to new conditions in the industrialised countries which will oblige developing countries, for economic and environmental reasons, to export semi-processed minerals?

Working Session III

Perspectives from Developing Countries on Environmental Policies
(With an introduction by Prof. Bharat Dhar, Benares, Hindu University)

While many developing countries are adopting measures to protect the environment and safeguard conservation, the pace is often slow because of poor awareness, high costs involved in mitigation measures and non-availability of cheap and appropriate technology and lack of appropriate guidelines for environmental protection and proper enforcement of laws.

In reviewing environmental policy development in developing countries, Prof. Dhar invited participants to assess:

1. Countries' capabilities to enact environmental legislation in terms of trained manpower, environmentally enlightened judiciary, the population and political will.
2. Effective administrative procedures for environmental control and monitoring.
3. Contribution/compensation by industrialised countries on developing countries in terms of:
   - direct financial assistance
   - help to develop indigenous technologies/methods in mining to mitigate environmental impacts.

Wednesday, June 26, 1991

Working Session IV

Economic Aspects of Environmental Protection in Developing Countries
(Introduction by Ms Beatrice Labonne, Chief, Mineral Resources Branch, UN/DTCD)

This session focused on comparative costs of environmental protection, impact of costs on international competitiveness, location of mining investment and trading patterns, and economic and financial considerations favouring and influencing environmental aspects of mining.

Participants assessed:

1. What are the trade-offs developing countries must make between the costs of environmental protection and the benefits of a mining project and how can these costs be estimated when social and economic growth is at stake? Is the cost of environmental protection prohibitive for developing countries?
2. How is the cost of environmental protection perceived in the developing countries?
3. What is the likelihood of new investments actually taking place in developing countries? Are investors willing? Are the countries concerned willing to make accommodations to their policies in areas other than environmental policy in order to attract investment?
4. Is it desirable to have mines operate under different standards according to the absorption capacity of the local ecosystem?
5. What might be the role of international organisations in assisting developing countries to balance technology and costs?

Working Session V

Tax treatment on Environmental Protection measures in selected developed countries
(Introduction by Mr Richard Axel Westin, University of Houston Law Center, Texas)

This session was devoted to surveying tax issues as they relate to mining enterprises in developing countries. In discussing microeconomic theories of taxation designed to address environmental policies as well as tax policies, the need to co-ordinate environmental and tax policy-making was addressed.

In discussing these issues, participants examined:

1. Does any country co-ordinate its tax and environmental policies?
2. What measures can be appropriately taken to ensure that the cost of workers' health losses is covered by the mining operator?
3. Should environmental taxes be remitted at the border, for example, when ingots are shipped overseas?
4. Should current deductions be allowed for future clean-up and mine closing expenses?
5. What interest rate should one use in discounting for future events?

Working Session VI

International Law and National Legislation, Regulation and Contractual Provisions concerning Environmental Protection of Mining
(With an introduction by Dr Thomas Wälde, Centre for Petroleum and Mineral Law Studies, University of Dundee, and Mr Lawrence McDonnell, Environmental Law Center, University of Colorado)

This session focused on the main issues of environmental protection in mining, from general policy towards implementation by legal and fiscal financial instruments. The necessity of an integrated approach of all policy instruments was discussed as well as concepts of regulation.

Important issues included:
1. Social interest objectives to be safeguarded.
2. Use of market in connection with regulation to achieve objectives.
3. Difference of command legislation with regulation of market activities.
4. Development of environmental standards by competition, emulation and self-propelled technological growth as by response to external pressures.

In reviewing these issues, participants discussed:
1. Are environmental legislation and guidelines going to divide the world into two categories: the countries with lax mining codes and those with more stringent ones?
2. Mindful of the difficulties faced by some developing countries in enforcing existing codes, how might they be expected to enforce environmental legislation?

Thursday, June 27, 1991

Working Session VII

Socio-Cultural Impact of Mining in Developing Countries
(With introductions by Mr Allen Clark, East-West Centre, and Ms Lorraine Ruffing, UNCTC)

Because of the enclave nature of most mining ventures in developing countries, any new mining operation will cause some degree of disruption to society. A new mine may employ several hundred, or even thousand, people and these people must be housed, clothed, fed, transported and taught new skills. A mine will spawn a series of service industries in the mine area, providing opportunities for local entrepreneurs. People inhabiting areas around a new mine have to be properly prepared for these changes in circumstances.

People residing in the development area must be relocated. Rapid build-up of a construction force (possibly of mixed national or ethnic origins) may create cultural tensions and strains on traditional support systems. Local inhabitants, lacking the necessary skills, may resent their exclusion from key jobs and resent the presence of better-paid expatriates.

Furthermore, careful planning is needed to maintain some means of social support if the mine is decommissioned, leaving segments of the local population unable to revert to their traditional way of life.

In reviewing the above issues, participants recommended guidelines to be followed to minimise the negative socio-cultural impact of mining on rural populations. The session also addressed artisanal mining and its impact on the environment.

Working Session VIII

Technologies and Management for 'Optimal Amount' of Environmental Protection in Mining
(With introductions by Mr Roderick Eggert, Colorado School of Mines, and Dr Alyson Warhurst, OECD Mining and Environmental Policy Project, Sussex)

The term 'optimal amount' of environmental impact has been chosen over the term 'low environmental impact' as a way to emphasise that environmental protection comes at a cost and that this cost must be commensurate with the expected benefits. Optimally, realistic goals should be pursued rather than striving to eliminate pollution altogether.

Participants discussed:

1. What role should national governments and international agencies play in developing/diffusing new technologies that reduce pollution by the extractive mineral industries?
2. Should governments and international organisations place relatively more emphasis on the development of new technologies or on the diffusion of existing technologies?
3. Should governments and international organisations place relatively more emphasis on creating the right business environment for encouraging technological development and diffusion by non-governmental organisations? Or on targeting specific technologies?
4. What roles relative to one another should the following policy tools play?
   - public enterprise
   - financial support for research in science and engineering (either through government-run research laboratories or grants to university and industry researchers)
   - education in science and engineering
   - information (i.e. government provision of data bases, advisory services, etc.)
   - financial and tax incentives (e.g. loans, tax credits, cost-sharing, or subsidies to encourage development and diffusion of new technologies)
   - patents.

Working Session IX

International Co-operation
(With introduction by Ms Ayesha Días, UN/DTCD)
This session reviewed the role of international co-operation agencies (both multilateral and bilateral) in helping developing countries to meet the environmental challenges for management of their resources. It also sought to recommend how technical co-operation should be undertaken to improve the effectiveness and impact on the environment of technical assistance projects.

Friday, June 28, 1991

Panel Discussion

With senior policy-makers and mining company executives.

Chaired by Ernst Ulrich von Weizäcker, Institute for European Environmental Policy, Bonn.

APPENDIX D

THE ENVIRONMENTAL IMPACT OF MINING (Case Studies)

1 INTRODUCTION

At the Berlin Round Table, Mining Journal Research Services presented a 100 page report on 'Mining and the Environment'. This report focused upon eleven case studies drawn from mining operations around the world. Six of these examples are repeated in detail at the end of this appendix.

As an introduction to the report, Dr Chris Hinde of Mining Journal Ltd noted that the environmental damage caused by mining must be set in context with the industry's benefits. These could be summarised as mining's three 'I's:

— Industrial: the need for metals and minerals in an industrialised world.
— Infrastructural: mining has a particular part to play in the infrastructure of developing countries.
— International: often overlooked, but mining plays an important part in the lubrication of trade between nations.

Industrial Role: Much of what we rely upon in day-to-day life is constructed of, or consumes, materials which are hewn from Mother Earth. This is not to deny the increasing relevance of:

thinning in metal sheet, etc
substitution of man-made materials
recycling
increased efficiency of utilization, and
alternative energy sources.

Nevertheless, for the foreseeable future there is a need for metals and minerals; indeed demand is likely to continue rising in the short term for most commodities.

Infrastructure: Mining has played an invaluable role in the recent economic development of many African, Asian and South American countries. Indeed it played a not insignificant part in the industrialization of Europe. This contribution comes in two parts.

First, the direct influence of the roads, power lines, ports, etc, necessary for the winning of ore. Other industries which were too marginal to justify this level of expenditure could then successfully flourish in the shadow of shaft towers and spoil heaps. Expenditure on exploration will also directly contribute to the wealth of a country through the use of local employment and equipment. The proportion of exploration dollars spent in the developing countries is some 25%, and the proportion is still increasing.

Second, the export of these primary and secondary commodities contributes to economic wealth through taxes, levies, etc, the two-way flow of material and equipment also provides the framework for other deals.

International: An increase in trade barriers seems certain; there is already a concentration of world trade in three blocs – the European Community, the U.S. and Japan. By the end of this second millenium the World will be encompassed by huge trade blocs which approximate to five continents:

- The largest will be the United States of Europe.
- North America is expected to overcome the difficulty of accepting Mexico into its 'free trade' club.
- South America will see a combination of the 'Andean Common Market' and the 'Southern Market'. These nine countries are likely to be joined by Chile.
- Post Apartheid South Africa will be dominant amongst the 'Southern Confederation' of that continent.
- Asia will have a stronger version of its 'East Asian Economic Grouping'.

Despite all the rhetoric, there is essentially little point in having an economic community unless it is to the benefit of those inside the club. Each trading bloc will insist on equality and fair trade. Barriers are likely to fall to the lowest common denominator – and an environmental tax is certain to be one of the more important components for the mining industry.

However, international trade in metals will not fade away because of higher taxes and bureaucracy. It *cannot* fade away – mining does not have the luxury, which is enjoyed by other industries, of choosing where to produce. The mining industry must slavishly follow the deposits. This trade will be especially important to the European Community, whose domestic mines produce only a fraction of the total fabrication requirements.

Into this cosy picture of mining's role in the international community a grenade has been lobbed; in fact a green time bomb. There is no denying the importance of the environment, neither is there any point in denying that the mining industry has an appalling record. However, many of the past misdemeanours were perpetrated through ignorance rather than malice.

At the Western Gold Show in November 1990, Amax Gold's Mr Tim Haddon noted that "the environment has not, historically, been of major concern to miners". This is a classic understatement! Indeed the words 'green' and 'environment' are noticeable by their absence from the Mining Journals of 150, 100 and 50 years ago. Even five years ago this topic received minimal attention. Now, almost every other item seems to refer to the issue, and Mining Journal carries a twice-yearly special supplement on mining and the environment. This switch reflects the sea-change in sentiment in the mining industry.

2 MINING AND THE ENVIRONMENT

What then are the areas which need to be considered when discussing mining and the environment? Many of the issues can be grouped under the simple headings below (the four 'I's):

- Issues — at its most simplistic, the different impacts on the environment caused by *old* and *new* mines/smelters.
- Impact — are we essentially talking of *short-* or *long-term* effects, and are the problems *visible* or *invisible?*
- Instigators — who has caused the problem; is it large corporations, small companies, or individuals?
- Innovators — who is doing what to put the situation right? Should it be left to the companies themselves, public pressure groups, or international agencies?

Issues: The whole industry can be split simplistically in half:
- existing operations, with mine designs and metallurgical processes developed at a time when the environment was of far less concern.
- recently completed or planned operations. These have the opportunity to be constructed using the latest techniques for pollution control and environmental sensitivity.

It is important to distinguish these two parts of the industry. The former will inevitably find it harder to adapt, and any changes will invariably be at increased cost. On the other hand, the more recent operations can be more easily improved, and may even become more profitable as a result of better control of waste material, dust collection, etc.

Impact: One method of categorising the impact of mining on the environment is to examine the time-scale of the disturbance; are we talking of short- or long-term disruption. Alternatively, we could subdivide the impact simply into visible or invisible disturbances. In the visible category could be grouped land subsidence, land denudation, and eyesores such as waste heaps. Invisible impacts of mining could include pollution of air and water, and radiation.

Instigators: The third area which needs to be considered when examining 'mining and the environment' is who are the instigators of the pollution? Again, on a simplistic level, we can divide the parties responsible into:
- large corporations
- small companies, and
- individual miners.

The increasingly onerous demands on the mining industry, in terms of both cost and procedures, is likely to benefit the larger organisations. Only the well-funded and diversified corporations will be able to devote the necessary time and money for research and permitting which is becoming so necessary. On a tonne for tonne basis, it is likely that the thousands of garimpeiro miners in Brazil and the Philippines, for example, do far more damage because of their unregulated operations than any of the major mining houses.

Innovators: Lastly, any debate on mining's impact on the environment should address who is, or should be, responsible for making the necessary innovations. Under this subject can be included the necessary motivation, funding and expertise to effect a solution. Again we have three groups:

- Private companies
- Public bodies, and
- International agencies.

Essentially this choice boils down to: do you trust the market place? For the supply of funding and expertise for new operations, the private companies can probably be relied upon to do what is necessary. However, where the mines and smelters are old, or even abandoned, it will be necessary to rely upon local public bodies and international agencies for assistance.

It is in this area where the International Council on Mining and the Environment has so much to offer. ICME was formed at the beginning of 1991 through the offices of the Mining Association of Canada and, in particular, its president Mr Keith Hendrick of Noranda. This body is looking to raise the necessary funding from the leading mining groups to finance programmes of education and research.

3 CASE STUDIES

3.1 Carajas Iron Ore Project, Brazil
3.2 Bauxite and Alumina, Jamaica
3.3 Revegetation of Tin Mine Tailings, Malaysia
3.4 Rössing Uranium Mine, Namibia
3.5 Mercury Hazards, Philippines and Brazil
3.6 National Environmental Strategies, Poland

3.1 Carajas Iron Ore Project, Brazil

This case study reports on the impact that a large-scale mining project can have on its environment, and on the measures being taken by the operator to ensure that these impacts are minimised. The source material for the study was provided by Dr Claudio Margueron, consulting minerals economist, Rio de Janeiro.

3.1.1 Introduction

Brazil is the fifth largest nation in terms of area (8·5 million km^2), occupying 47% of the South American continent. It also has the ninth largest population, at approximately 150 million people. The country is rich in mineral resources, with estimates indicating that Brazil has over one half of global iron ore reserves. Apart from iron ore, it is amongst the largest producers of niobium, iron ore, manganese, quartz, beryl, mica, bauxite, tin, gemstones and tungsten, but has a significant resource gap in petroleum.

One of the major development projects undertaken during the 1980s was the bringing on stream of the Carajas iron ore deposits in the Amazonia region of

Brazil. The reserves in the Carajas Range are of the order of 18,000 Mt, and form the largest-known single high-grade (66% iron) iron ore deposit anywhere. The discovery of the Carajas iron ore deposits prompted further exploration for minerals, resulting in the discovery of manganese, nickel, copper, gold, bauxite and tin resources within the surrounding region.

The country's potential is gradually being realised, and Brazil anticipates that it will emerge as a significant world economic power by the end of this century. To do so will require the continuing development of its natural resources, the Carajas region forming one of the principal development centres in this context.

The Carajas deposits were discovered in the mid-1960s. The two principal iron ore deposits, the North and South Ranges, lie about 35 km apart. Production commenced in 1986 at a rate of 15 Mt/y, a figure that rose to 25 Mt in 1987. The projected annual production during the early 1990s is 35 Mt of ore.

Companhia Vale do Rio Doce (CVRD), which controls the Carajas iron ore project, is one of the largest and most diversified mining companies. Throughout the development of this and other natural resource projects, the company has acquired a reputation for its pioneering planning concepts that address the potential environmental impacts of a project at an early stage. While undertaking the development of major natural resources into world-class mining operations, the company is reported to consider environmental protection as an integral part of its sustained development programme.

The Carajas district is covered with almost unbroken equatorial rain-forest, characteristic of the Amazon basin region. The rain-forest is broken by clearings which correspond to the ore deposits, rising 300-400 m above the level of the surrounding land. The low, sparse vegetation in the clearings is due to the high iron content of the soil above the ore, which hinders the development of a denser vegetation cover.

The region is located some 550 km south of the Atlantic coastal port of Belem. The average annual rainfall is 1,900 mm, 90% of which falls in the period November to May. The natural population density is extremely low. Seven residential areas have been built to house the project's employees, the Carajas Residential District alone having a capacity for an initial 11,000 inhabitants. The other town sites are located at various intervals along the line of the railway.

The Carajas iron ore project has assisted in opening up the Amazonia basin. The railway has been constructed through areas that previously had limited transport capabilities. Export sales of iron ore from the project generate around $700 million annually in foreign revenue.

3.1.2 Environmental Impacts

The Carajas project is unusually large in its scale, and is one of the few recent examples of the minerals industry leading the way in the development of an entire region of a country. Thus, the project's environmental impacts are felt not only in the immediate surrounds of the mining area, but also directly and indirectly across a wide region of what was previously undeveloped country.

During the early stages of the Carajas iron ore project, a number of other ambitious projects were attempted in the Amazonia region. Most failed, principally because there was a complete lack of knowledge of the ecosystems of the region. Prior to the iron ore project, there was little interest, both regional and scientific, in the Carajas area. Agriculture in the region was primarily at subsistence level, with the main product being rice. Timber was cut, some gold was mined, and ceramics were made.

Encouraged by CVRD, Brazilian scientists commenced a series of environmental studies in the Carajas region. Research projects were undertaken on the flora, fauna, archaeology and water quality of the area, and a policy of continuous environmental monitoring was established.

With the completion of the Carajas railway, the region entered a period of rapid transformation. Many new villages and small towns were established along the line of the railway. CVRD recognised that the development brought about by the rail construction had upset the regional environment, and undertook a programme of urgent corrective measures.

The direct impacts on the regional environment, from the mining, handling, processing and transportation of the iron ore, include: soil erosion, waste pile slides, dust generation, water contamination, forest fires and tree felling, liquid wastes, and air pollution.

3.1.3 Response

The early 1980s saw the first appearance in Brazil of an increased awareness of the importance of the relationship between man and the natural world, and the consequent need to protect the environment. Another noteworthy aspect to emerge from the Carajas experience has been heightened ecological awareness among CVRD employees, which has assisted greatly in the identification of potential and actual problem areas.

After two years of intensive effort to minimise the effects of its activities on the environment, CVRD is currently working on various Environmental Protection Master-plans for implementation during the 1990s. These take into account various fast-moving developments in Brazilian environmental legislation, as well as demands made by society during the past few years.

CVRD has set up an Environmental Studies and Counselling Group, consisting of top-level scientists and technicians. This department draws up guidelines for the company's operational divisions, and is developing environmental programmes for the various areas of Brazil where the firm operates. In all, the past nine years have seen CVRD invest some $300 million in both environmental reconnaissance surveys and controlling the impact of its operations. A further $100 million is to be spent by 1993 in a four-pronged programme that will include the rehabilitation of damaged areas, and environmental control and research.

One aspect of the programme will be the maintenance and management of over 500,000 ha of forest in the Carajas Range and along the line of the Carajas railway. This will include the creation of ecological and economic zoning systems, together with the provision of medical and social services to the indigenous population in the Carajas project region.

Green belts are to be planted along the line of the railway, while on the operational side, additional measures to cover environmental control will be implemented. Soil erosion and conservation will also be covered, as well as further environmental studies into flora, fauna, water resources and alternative sources of energy.

The Carajas region is also benefitting from Federal action. The Eastern Amazon Forestry Programme was recently implemented, having as its basic objective the encouragement of projects in the region that will contribute to the preservation, conservation and rational utilisation of natural resources. This pioneering strategy is aimed at reversing the environmental degradation suffered by the eastern Amazon region in the past decade. The programme was conceived to allow, through non-profit organisations such as the Brazil Amazonia Foundation, financial resources in industrialised nations to be applied to environmentally orientated projects in the eastern Amazon region. The target is to undertake reforestation of 1 Mha of land in the area of influence of the Carajas railway in the next 10 years, and two large-scale projects in eastern Amazonia are currently being analysed.

3.1.4 Assessment

The Carajas project has received great international attention in the past few years, more specifically as a side issue during the current debate on rain-forest destruction than on its own account. There can be no doubt that international pressure by governments and concerned organisations has had a significant effect on the way in which the environmental impact has been handled. In addition, increased public awareness of the rain-forest issue, together with reports of continuing ecological damage by agriculture, mining and logging in Brazil and elsewhere, has ensured that effective environmental action is taken by both CVRD and the Brazilian Government.

CVRD is now at the forefront of protecting Brazil's rich natural heritage. The Carajas iron ore project is seen as a showcase that puts into practice all that the company has learned during the 50 years that it has operated in the southern part of the country, centred on the Itabina-Vitoria Minas railway-Tubanão port system.

Consequently, the Carajas project is seen as the international state-of-the-art in the development of large-scale mining operations where environmental concerns are of paramount importance. It is estimated that CVRD has invested some $2,800 million in Carajas. Of this, $53 million were spent on environmental studies in the period 1980-1985, but success was only achieved because the company had both adequate funds and sufficient numbers of well-trained supervisory personnel to implement and monitor these environmental programmes.

CVRD fulfills all parts of the Brazilian national environmental regulatory code, and will have no difficulty in meeting any future requirements that may be introduced. Nevertheless, its attitudes and successes must be seen in the regional context, and in terms of the high level of international attention that will continue to be given to the Amazon basin.

3.2 Bauxite and Alumina, Jamaica

Bauxite is one of Jamaica's principal natural resources. In a country with a population density of around 200/km^2, it is inevitable that the industry will have an impact on those around it. Control measures are in force to minimise these impacts. The source material for this study was provided by Mr Godfrey Perkins, Permanent Secretary of the Jamaican Ministry of Mines and Energy, Kingston.

3.2.1 Introduction

The mining of bauxite and the subsequent production of alumina are key factors in Jamaica's economy. Responsible for more than 60% of the country's export earnings, the industry is, however, wholly dependent on international market trends and events in the world economy. The Jamaican Government is pursuing a number of objectives in an effort to ensure stability and development of this important area of the economy; one of these is the expansion of the country's alumina production capacity.

Despite the downturn in the international economy, 1990 proved to be one of the best years on record for the Jamaican bauxite industry. The country recorded its highest shipment of alumina, which at nearly 2·9 Mt represented a 34% increase over 1989 and the highest output since 1974. Bauxite production in the 1990-91 financial year amounted to 11·18 Mt, with raw bauxite exports totalling 3·9 Mt. The foreign exchange received for these exports amounted to approximately $US600 million.

Jamaica has bauxite reserves estimated at 1,500 Mt. The deposits occur as pockets within white limestones, and are often the sites of small subsistence farms and rural dwellings. The bauxite occurs in areas of low rainfall, and the high permeability of the underlying limestones gives rise to shortages of surface water. Most of the mines are in areas where those living nearby have little or nothing to do with their direct operations. Mine and plant personnel usually commute to work from other areas.

Jamaica is probably unique in having so many bauxite mines and alumina plants in so small an area. Open-pit mining competes for land with subsistence agriculture, which provides several food crops annually, as well as pasture land for raising cattle. In such a situation, the economic arguments often weigh in favour of mineral extraction taking precedence over agriculture, with the consequent need to find alternative land for the displaced farming communities.

A further associated problem is that the land, once mined, cannot readily be returned to agricultural use, since the opencast operations remove substantial volumes of bauxite ore. Even the necessity to store the red mud that results from the conversion of bauxite into alumina does not assist in this respect, since the mud cannot be stored in the mined-out workings.

Bauxite has been mined in Jamaica since the early 1940s. The first companies, Alcan and Reynolds, were later joined by Kaiser, and by 1957 these three North American companies owned an estimated 5% of the country's total land area. At that time, Jamaica was the world's largest producer of bauxite. These companies were later joined by Alcoa, the Alpart consortium (Anaconda, Kaiser and Reynolds), and Revere.

During the 1970s and 1980s, the situation changed considerably. The Revere alumina plant closed in 1975, and Reynolds ceased operating in Jamaica in 1984. New legislation brought bauxite reserve land into state ownership, with the government granting mining leases to the former land holders to cover 40 years of future production.

In 1985, Alcoa closed its Clarendon alumina plant, citing the depressed world market as the principal cause. The plant was reopened by Clarendon Alumina Production Ltd., which had been set up by the government for that purpose. The Alpart alumina refinery was also closed in 1985, and the government later revoked the consortium's mining lease. In 1988, the Clarendon plant became an equally owned joint venture between Clarendon Alumina Production and Alcoa (Jamalco), with agreement between the parties on a variety of issues, including the construction of a new red mud lake and a reduction in the bauxite levy paid by Alcoa.

Today, the Jamalco, Alpart (reopened in 1989) and Alcan alumina refineries are in operation. All three companies have indicated their intention to produce at maximum capacity. In addition, the Jamalco and Alpart plants are considering proposals to increase capacity by some 500,000 t/y over the next three years, and a feasibility study is scheduled for a new greenfield site refinery on the north coast of the island. Jamaica is now the third largest bauxite producer, after Australia and Guinea.

Current government policy is to ensure that exports of alumina are increased in preference to exports of raw bauxite. The reason for this is economic – processed alumina is worth many times the value of bauxite. However, this requirement for expanded alumina refining capacity will have the unavoidable environmental consequence that red mud storage capacity will have to be increased in proportion. Each tonne of alumina requires the production of an equal amount of red mud, and in 1976, a government investigating committee reported that properly constructed red mud lakes require about 50 ha of land per alumina plant every five years.

Jamaican bauxite mining is an exclusively opencast operation; there are no underground operations. The mining process involves stripping overburden and then recovering the bauxite, which is transported to the drying and refining plants for treatment. After drying or conversion to alumina, the products are transported to the coast, and are shipped for export from dedicated loading facilities.

3.2.2 Environmental Impact

With such a high density of population in close proximity to the bauxite mining and treatment operations, it is inevitable that the industry will have an impact on the surrounding districts. Three main areas of impact occur; population displacement, dust and noise, and red mud disposal.

The importance of bauxite and alumina within the Jamaican economy dictates that mining takes precedence over subsistence agriculture in the use of land. This policy therefore necessitates the relocation of those people living in a proposed mining area, at the expense of the company involved.

A survey undertaken in the early 1980s gave an indication of the type of population typically involved in this situation. Over half were farmers, with an average farm size of 2·7 ha, which is sufficient to feed a typical family. Outright purchase of the land was the normal type of transaction between the mining companies and the farmers, although this system was later modified to include combinations of cash purchase, land exchange and resettlement.

In some cases, entire communities have been resettled. An example is the community of Bowens, near the Jamalco Clarendon plant, which is currently being relocated complete with its houses, church and school in a move to address certain environmental issues and to release land for the disposal of bauxite waste.

The second area of impact is emissions. Jamaican bauxite is of extremely fine particle size, as is the calcined alumina. Dust can escape from most stages of the operations, including mining, drying, calcining, transportation and shiploading. Neither bauxite nor alumina dust is considered hazardous to health, but both constitute a serious nuisance in environmental terms.

Bauxite dust is red, while alumina is white. Dust fallout in populated areas gives rise to additional cleaning requirements, while it also settles in the rainfall catchment tanks, which for many provide the only source of domestic water.

Noise from blasting and the operation of heavy mining equipment can also be intrusive, as are vibrations and airblast overpressures from blasting operations. However, the most widespread source of complaints has proved to be the accelerated corrosion of galvanised steel house roofs from acidic sulphur dioxide emissions resulting from the use of high-sulphur (2%) heavy fuel oil in the alumina plants. These effects have been recorded over distances of up to 6km from the alumina plants, although the matter is complicated by the presence of other facilities burning fuel in the same areas.

Overall, the factor of greatest impact on the environment is red mud. The provision of adequate red mud storage facilities is critical to the continuing viability of the Jamaican bauxite industry, and recent years have seen continuing research towards new methods of red mud disposal. According to information presented at a 1986 conference on bauxite waste disposal, between 1952 and 1985 Jamaica produced 159 Mt of red mud from its bauxite processing operations.

The most commonly used method of storage has been containment of the fluid mud in sealed lagoons. Mined-out bauxite pits cannot normally be used in this context, since groundwater must be protected from contamination through seepage. This means that storage for the red mud has to be created on what might otherwise be productive agricultural land, and the availability of suitable storage sites remains a problem for the companies concerned.

Red mud presents both physical and chemical problems. Its sludge-like properties necessitate proper containment facilities, while it is also chemically caustic, thereby inhibiting attempts to establish viable revegetation programmes. Its consistency enhances moisture retention in the mass, while the surface can become baked and cracked during dry periods. It has few uses, although research is continuing into developing it as a raw material for construction materials.

3.2.3 Response

Faced with having to operate near populated areas, the Jamaican bauxite industry is not in a position realistically to avoid environmental impacts, but rather to control them to acceptable levels. This in turn can lead to conflict, since the mining companies and the largely rural population so affected can have significantly different perceptions of control.

The problem of land acquisition and resettlement has been addressed through the introduction of a range of options, as outlined above. This area of impact will continue, and is inevitable in the circumstances. One benefit to the displaced population is that the standard of replacement housing provided is often of a higher standard, an important consideration in a situation where a farmer's entire assets may be his house and his land.

The success with which these policies have been implemented appears to have varied with the company concerned. The study into social issues raised by the land transactions reported distinct differences in both the initial and long-term attitudes of those concerned, depending upon which company had been involved with the purchase of the land. This in turn can affect the overall perceptions of the local population towards the mining companies. A major factor in this area was the nature of the transaction; those paid only in cash showed a higher rate of dissatisfaction with the transaction than did those who were either allocated land or were rehoused.

Visual impacts such as dust and corrosion are also impossible to eliminate, but can be minimised. Much depends on good management and maintenance, and it is significant that Kaiser's Port Rhodes operation, which once boasted high compatibility with its surroundings, is now the source of repeated dust nuisance complaints because of a fall-off in the maintenance of its systems. This has probably been a result of reduced expenditure due in turn to economic pressure on the company, which has been instructed by the government to improve the situation.

The problem of roof corrosion is harder to solve, both the mining companies and the Jamaica Bauxite Institute having undertaken field studies on it. The companies have been making *ex-gratia* payments or other arrangements to replace galvanised iron roofs with aluminium sheet, which is more expensive. In extreme cases, such as the Bowen example noted above, entire communities have been relocated away from the area of influence of an alumina plant.

Red mud will continue to present storage problems until some alternative use can be found for it. Conflicts have arisen through the purchase of land for red mud storage, and in the short term the only solution appears to be the adoption of new technology aimed at improving its handling and storage characteristics.

A new system of red mud storage, based on deep thickener technology, has been developed in Jamaica by Alcan, and has been in use at the company's Ewarton plant for some years. It has proved so successful that it is now being licensed by other bauxite mining companies both in Jamaica and overseas.

3.2.4 Assessment

The significance of bauxite and alumina to Jamaican foreign earnings exceeds that of tourism and agricultural exports. Moreover, the state of the country's economy, heavily indebted and set back by the effects of a major hurricane in 1988, has dictated that Jamaica should proceed down the route of adding value to its raw materials, by increasing alumina plant capacity, and hence the amount of red mud that the industry will generate.

The country faces a difficult dilemma. Other bauxite producers have the advantage of large land areas and sparse populations, of which Jamaica has neither. In addition, the country has to balance the needs of industry against those of its agricultural exports and tourism, both of which are permanently important contributors to foreign earnings.

The Jamaican bauxite and alumina industry has passed through some uncertain times, but now appears to be in a buoyant position. The question remains as to how far the country should rely on one sector, given the vagaries of the commodity markets, but in the foreseeable future at least, it seems set to optimise bauxite and alumina income.

In many respects, the social impacts of this industry are not unique to Jamaica, but can be seen in many developing countries where a rural population has to co-exist with industrial projects. The prime concern is to minimise these impacts which, as has been shown, can often be achieved by careful planning, sustained maintenance and the use of appropriate technology.

3.3 Revegetation of Tin Mine Tailings, Malaysia

This case study describes one possible route for mine tailings rehabilitation, using as its example the planting of oil palm trees on tin mine tailings in Malaysia. Both industries are significant to the Malaysian economy. The source material for this study was provided by Tuan Haji Abdullah bin Yusof, consultant, Kuala Lumpur.

3.3.1 Introduction

Alluvial tin production was a mainstay of the Malaysian economy for many years. Until recently, the country was the world's leading tin producer, with both on- and off-shore mining operations, and supporting smelting facilities. It is also a major tin metal trading centre.

During the peak period of tin production, in the mid-1970s, Malaysian mines produced up to 76,000 t/y of tin-in-concentrates. Today, the country's production has fallen significantly, to some 29,000 t/y, largely as a result of the collapse in 1985 of the International Tin Council's pricing agreement, and the substantial overhang of tin stocks that subsequently depressed the commodity market. In the late 1980s, tin production accounted for around 2% of Malaysia's export income, reflecting the dramatic decline in the number of operating mines that took place in the late 1980s.

Historically, the other principal Malaysian raw commodity has been rubber. In 1980, the country held a 40% share of world natural rubber production, but while production levels have remained relatively constant in recent years, the country's share of the market has fallen with the expansion of other producers.

Faced with declining revenues from its natural products, Malaysia has undertaken a major reappraisal of its economic base. Diversification has been encouraged through the introduction in 1970 of the New Economic Policy, which has led to the introduction of both new industries and new cash crops in an effort to reduce the country's dependence on the vagaries of the international commodity markets. As a consequence, Malaysia now enjoys an expanding and diversified economy that is one of the most buoyant of the Pacific Rim nations.

One of the most successful expansions has been in oil palm cultivation. Annual palm oil production has increased from below 3 Mt in the early 1980s to its current level of around 5·5 Mt, placing Malaysia first in world palm oil production. Today, the country has some 1.85 Mha under oil palm cultivation. Palm oil is a versatile vegetable oil, with uses in foodstuffs, cosmetics, soaps, and in the rubber and other industries.

3.3.2 Environmental Impact

Gopeng Bhd. is one of the oldest mining companies in Malaysia. It commenced operations in 1900, and has operated continuously ever since over an area of some 5,000 ha. It currently operates a large opencast mine near the town of Gopeng in the State of Perak. The mine is currently nearing the end of its life, although annual production remains at around 1,200 t of tin-in-concentrates. Mining also provides the main employment in the district, which has a population of some 50,000 people.

The mining operation consists of stripping overburden, removal of the tin-bearing gravels beneath, and then replacing the tailings from the tin recovery plant in the void created as the mining face progresses forward. Both coarse and fine tailings are generated by the mining and recovery processes, exhibiting different particle size and other characteristics.

The legal framework currently in force in Malaysia requires mining companies to 'fill and level' mined-out areas. There is, however, no provision for rehabilitation of old opencasts or tailings areas, and ex-mining land is often used for the construction of low-cost housing and for the development of industrial sites.

Malaysia now has an estimated 200,000 ha of both sand and slime tailings throughout the old on-shore tin mining districts. The potential for utilising this land for generating both revenue and employment is considerable, and while the government has recognised this potential, a comprehensive system of rehabilitation and re-use has yet to be fully implemented.

Throughout its life, Gopeng Bhd. has actively engaged in agriculture on the lands surrounding its mining operations. This has included oil palm cultivation, and with the company facing limited remaining mining reserves — and hence reducing profits from mining — a logical step appeared to be to investigate the economics of planting oil palms on areas of land that had been worked out, and were in need of rehabilitation.

The company commenced experimenting on trial tailings plots in 1982. The initial plots were part of a 200 ha area that had been covered with tailings and for which there was no further prospect of mining. Two types of area were available, pure sand tailings and sand tailings that had been mixed with slimes. Of these, the latter material has better water retention characteristics.

The use of tailings for agriculture was not new, however, since another company, Berjuntai Tin, had previously experimented with rubber trees on slimes tailings at its operations at Batang Berjuntai in the State of Selangor. In addition, tapioca had been widely grown on tailings material, as had fruits such as guavas, pomelos and star fruit, and cash crops such as asparagus.

3.3.3 Response

Gopeng Bhd.'s initial oil palm experiment was carried out on a 3 ha slimes tailing trial plot. In 1985, the experiment was widened to 3 ha of slimes tailings and 4 ha of sand tailings. Both were planted with seedling oil palm trees, which were obtained from the company's own nursery. With encouraging initial results, 1988 saw a further enlargement of the scheme, with three blocks of 10 ha each being set up to run comparison trials between the soil conditions found on sand tailings, slimes tailings and virgin land.

The two types of tailings generally exhibit significantly different characteristics. The sand tailings often require initial levelling, while the slime tailings are already flat. In addition, the sandy material has little organic content, making it more difficult to establish vegetation, has low water retention capabilities, and has high surface soil temperatures and evaporation rates. By contrast, the slimes tailings provide a better medium for vegetation establishment, but can present problems with both water-logging and the creation of hard pan conditions which inhibit water percolation.

Slimes Tailings: Seedlings planted in September 1985 showed signs of water stress until the hard surface of the ground was broken up by repeated harrowing. This promoted the passage of water from the semi-dry surface to the area of the root systems, and also permitted the planting of a cover crop, designed to prevent excess evaporation and to add organic material to the depleted surface layer. A regular scheme of applying fertiliser to the plants was adopted, and the trial showed that oil palms could indeed be introduced satisfactorily on this type of tailings material.

Sand Tailings: The seedlings planted on the sandy tailings plot did not perform as well as those planted on slimes tailings. The seedlings required mulching with empty fruit bunches and fibre obtained from the nearby palm oil mill in order to retain moisture and to cool the ground surface. A cover crop was used to bind the sandy soil and to provide essential nutrients to the surface region. Fertilisers were also applied, and although plant growth throughout the trial plot was found to be uniform, there were wide variations in leaf colour between trees planted in areas where there were both sand and slimes intermixed, and areas where the growing medium was wholly sand.

3.3.4 Assessment

The most recent phase of the experiment has been the comparison of production results obtained from oil palm plantations established on different soil mediums. The same planting and maintenance regime was used on all three 10 ha blocks, which were set up on sand, slimes and virgin land. The results obtained, in terms of yield of fresh oil palm fruit bunches per hectare, are tabulated. A yield above 20 t/ha is considered good.

Type of land	*Months in harvesting*	*Yield (t/ha)*
Virgin land	39	24·50
Virgin land	37	20·70
Former slimes area	38	28·00
Virgin land	35	15·85
Sand tailings	28	16·68
Virgin land	20	18·27
Former slimes area	19	21·79
Virgin land	20	13·75

Based on the initial results from these experimental plots, as well as analyses of the soils, it has been concluded that oil palms are a suitable crop to be planted on both sand and slime tin tailings, and that the expected yields of between 18 and 20 t of fresh fruit bunch per hectare would result in a profitable operation. It is also expected that yields would vary with the variability in the types of soil. In general, it was found that palms planted on clayey soils exhibited better growth characteristics than those planted on sandy soils, while the application of fertilisers needed to be undertaken on the basis of low dosages at frequent intervals in order to minimise losses through leaching.

Other factors, applicable to the establishment of crops on any type of depleted soil (and not just to the Malaysian conditions), include the need for proper water table management, so that plants do not suffer from water stress, and the necessity for planting ground cover species that will both bind the surface of the soil and assist in the re-establishment of nutrients and soil structures lost during the mining process.

The experiments conducted by Gopeng Bhd. have implications that extend significantly wider than the immediate situation in peninsular Malaysia. Many countries have areas of unreclaimed mine workings, and knowledge gained in their successful rehabilitation in one country may be transferred to assist in similar projects elsewhere. In the Malaysian context, there are some 200,000 ha of tin tailings in the Peninsula, of which about 80% are the more problematic sandy tailings. The concept of rehabilitating currently under-utilised land to raise much-needed cash crops has great potential.

While the cost of undertaking experiments in both rubber and oil palm cultivation on tin tailings has up to now been borne by mining companies, an expanded scheme would complement the Malaysian Government's environmental rehabilitation policies. It is interesting to note that the impetus for these experiments has come from mining companies seeking to replace falling revenues. In so doing, they have made a significant contribution to the development of reclamation technology.

3.4 Rössing Uranium Mine, Namibia

Uranium mining and recovery present particular environmental concerns. This case study reports on the way in which the Rössing uranium mine in Namibia handles various unusual environmental aspects that arise from the combination of its location and its product. The source material for the study was provided by the RTZ Corporation and its subsidiary, Rössing Uranium.

3.4.1 Introduction

Namibia is a newly independent nation. The country has significant mineral resources, which supply a substantial proportion of its foreign earnings. Included in these resources are the Consolidated Diamond Mines' operations, recovering alluvial diamonds from palaeoplacer deposits, and the Rössing uranium mine, one of the largest open-pit operations in Southern Africa.

Namibia has an area of over 800,000 km^2 (more than twice the area of Germany at 357,000 km^2), with a population of less than 1·5 million. The 1,300 km-long South Atlantic coastline is dominated by the Namib desert, 80-120 km in width, while in the east the country shares the Kalahari desert with neighbouring Botswana.

Owing in part to the harsh desert climate, the population density of Namibia is sparse, with significant populations only in the far north of the country, bordering Angola, and in the principal towns and cities. The national economy has traditionally been supported by three primary industries – fishing, agriculture and mining. In recent years, however, increased stress has been seen on the first two of these, with drought and declining fish stocks placing pressure on the minerals sector to provide an increasing proportion of national revenue.

The scale of Rössing's importance to the fragile Namibian economy cannot be underestimated. The mine's sales reportedly account for over one-third of total Namibian merchandise exports, and (on average over recent years) Rössing has contributed some 13% of the national Gross Domestic Product. During 1989, the mine contributed over R400 million (approximately $160 million) directly to the national economy through the payment of taxes and salaries, and through the local purchasing of goods and services. In recent years, the mine has supplied an estimated 45% of all income tax collected from the business sector and, overall, Rössing provides about 18% of total locally generated government revenues.

Rössing is located in the Namib desert, 65 km northeast of the coastal town of Swakopmund. In such a location, water supply is a critical parameter, the mine being provided with its requirements through the Central Namib State Water Scheme, which is capable of supplying local consumers — including Rössing — with up to 15 Mm^3/y of water via holding reservoirs at Swakopmund.

The impact of a large-scale operation such as Rössing's on a fragile desert environment is not limited only to water usage. Dust generation is of particular significance from both the points of view of nuisance and radiological hazard, and the mine undertakes a stringent environmental protection programme, designed to prevent unnecessary risk to both the locality and the individual.

3.4.2 Environmental Impact

Rössing operates an open-pit mine and uranium recovery plant. The deposit in which the uranium is found is unusual in terms of its economic geology, being low-grade and hosted in granite, rather than the more common sedimentary host rocks. It was delineated in the late 1960s, and production started in 1976; however, it was not until 1978 that the full design capacity of 4,540 t/y of uranium oxide product was achieved. Although the confidential nature of the uranium sector means that throughputs, reserves and grades are not released by the company, it is recorded that during 1990 the recovery plant had an average throughput of 34,000 t/d, indicating an annual ore throughput of around 12 Mt. Similarly, estimates of ore grades generally fall into the range 0.03-0.04% U_3O_8, making Rössing a low-grade operation by international standards, and thus necessitating the high ore processing rate. The mine is believed to be operating at about 80% of its design capacity.

The Rössing deposit was brought into production, and is now operated, by a joint venture between RTZ (46·5%), Total Cie. Minière (10%), the Industrial Development Corp. of South Africa (10%), Uranges (5%), Gencor (2·3%), the Namibian Government (3·5%) and others. The operating company is Rössing Uranium Ltd.

Mining in a desert area has different environmental impacts than those normally encountered in other climatic conditions. Often the ecology of the region is fragile, dependent on scarce water resources and finely adapted to survival in a harsh climate. In the case of Rössing, the potential areas of impact are magnified and diversified for a number of reasons, not least of which are the scale of the operation and the nature of the product.

Water: Large-scale mining operations can have high water consumption requirements, particularly where a high degree of purity is demanded in a chemical extraction process. Rössing has no immediate water supply, relying instead on the State reticulation system that provides water for the entire Central Namib region. During the 1980s, Rössing's water consumption averaged 30,000 m^3/d; recycling commenced in 1981, and by the end of the 1980s, up to 40% of daily water demand was being satisfied by the use of water recycled from within the operation itself.

Dust: This is a problem in any arid-area mining operation. At Rössing, the problem of the visual nuisance is compounded by the potential for physiological damage from both silica and radioactive materials. Areas of dust generation include the open pit itself, the haulage roads on the surface, the processing plant, the waste rock dumps and the tailings impoundment dam.

Groundwater Contamination: The tailings dam is a potential source of groundwater contamination, since the effluent from the treatment and recovery plants contains acids, metals and unrecovered uranium. While much of this material is held either within or immediately below the dam, the effluent from the dam tends to liberate nitrates and other salts held in the underlying soils. However, there is no industrial or domestic water supply capability near Rössing, so that any long-distance seepage that evades the mine's collection system is deemed unlikely to have any significant environmental effects.

Radiological Effects: The environmental concern that is unique to this type of operation is the effects of radioactivity. The Rössing mine produces a low-grade ore which, however, still has the ability to emit alpha, beta and gamma particles. Only the first of these is considered a health risk to the individual, and personnel at risk of exposure to alpha radiation are provided with filtration equipment. In an underground mine, radon emissions from the orebody could also present a hazard, but in the open pit environment, dilution is rapid, and radon is not perceived as an environmental problem at Rössing.

3.4.3 Response

Rössing has maintained stringent environmental controls on its operations. These are seen as being essential in both economic and environmental protection terms, particularly in the cases of water and radiation. The company is safety-conscious, and reports that to date no employee has been found to have a medical condition resultant upon either exposure to radiation or to silica dust inhalation.

Water: The company has developed a widespread water recycling programme, designed to minimise the mine's fresh water requirements. Some 80,000 m^3/d of tailings are discharged to the tailings dam, containing an estimated 16,000-20,000 m^3 of water. After settling, much of this water (*c.* 12,000 m^3/d) is recycled for uses such as the milling section of the processing plant, haulage road dust control and dust control within the open pit.

Dust: Airborne dust is minimised through the regular use of water spray trucks on the mine haulage roads. In addition, all drilling in the mine is wet, and all blasted rock is wetted using water cannon trucks before and during loading. Water sprays are used to minimise dust generation when trucks tip ore into the primary crushers. Some 2,000-2,500 m^3/d of water is used in this way for dust supression. Of this, an average of 38% is used for rock pile wetting.

This policy is undertaken for two reasons; first to control respirable silica, and secondly to prevent alpha radiation reaching internal tissue. From the point of view of the individual, all mobile equipment operating in the open pit is equipped with filtration modules, so protecting the machine operators, while in those areas where either dust or radiological hazards could occur and enclosure of the individual is not possible, suitable individual protection equipment is provided.

It is not considered necessary to spray water on the waste rock tips and the low-grade ore stockpile areas, since there is very low dust generation from these areas of the mine, with no noticeable effect on the surrounding environnment.

Rössing operates a comprehensive air quality surveillance programme, with air samplers located at various points on the mine and in the nearby residential areas. Typical dust levels are about 60 $\mu g/m^3$ at a distance of 2 km from the mine, and 30 $\mu g/m^3$ 12 km away.

Groundwater: Seepage from the tailings dam is minimised through the use of a system of wells and catchment trenches, which permit only a small proportion of the total seepage to enter the underground water flow beneath the bed of the nearby Khan River. Regular monitoring is undertaken to ensure that effluent quality controls are being maintained, the results being supplied to the national Department of Water Affairs. Water management is of critical economic importance to the operation, and a computerised seepage model has been created in an effort to predict seepage flows and to improve water management techniques.

Radiological Protection: The company reports that on average, a production worker at Rössing receives a total of around 3·0 mSv (milliSieverts) of radiation per year. This, it states, compares with a natural expected background radiation level of 1·6 mSv, and compares favourably with the internationally accepted maximum annual exposure limit of 50 mSv for workers in the nuclear industry. Workers are required to wear dosimeters, and to provide medical samples for regular checking. Full safety and protection equipment is provided for workers in higher risk areas, such as the uranium oxide packaging plant.

The company maintains a widespread monitoring system throughout the region around the mine, using computer modelling to simulate air dispersal and radiation characteristics. Regular surveys of both natural flora and fauna are undertaken, particularly along the Khan River. The only other significant gaseous emission from the operation is sulphur dioxide, produced during the production of sulphuric acid. Monitoring of SO_2 is undertaken on both personal and regional level, with modelling being undertaken to ensure that no SO_2 damage is caused to the environment.

Rehabilitation: At present, no measures are being taken to rehabilitate the waste rock tips. On the tailings, extensive studies have been undertaken to establish the best possible cover in the short-, medium- and long-term. Vegetation with natural species of salt bush is only considered a short-term possibility in an area of only

30 mm/y rainfall, but a 1·5 m-thick layer of coarse alluvial sand has proved both to be an effective radon barrier and to prevent wind erosion of the tailings.

3.4.4 Assessment

The Rössing operation is unusual for a number of reasons. It mines low-grade ore with prospective medical risks, and it is in a fragile desert environment. Coupled to this is the absence in Namibia of a comprehensive environmental legislative structure for the minerals industry.

However, the company considers that its standards and monitoring are in line with international practice, and is confident that these will meet comfortably the new Namibian environmental codes once these are introduced. Meanwhile, in circumstances where no Namibian standards currently exist, such as for SO_2 emission, Rössing Uranium operates to U.S. standards.

There can be little question that, within the framework of minerals projects in developing countries, Rössing has special features. Nevertheless, it has shown that it is possible to operate a large open-pit mine and processing plant in an environmentally compatible manner, while maintaining viability and contributing substantially to the national economy.

3.5 Mercury Hazards, Philippines and Brazil

This case study reports on the findings of research into mercury contamination resulting from gold mining in two developing countries. It also reviews the actions taken to control this use of mercury, and looks at possible alternative technologies that could replace it in the small-scale mining context. The source material for this study was provided by Dr Stefan Cramer, technical adviser in the Philippines.

3.5.1 Introduction

Mercury pollution from small-scale gold mining represents an environmental problem in developing countries. Worldwide, several hundred tonnes of mercury are released every year. Two examples of polluted river basins in tropical environments are the Agusan River in Mindanao (Philippines) and the Madeira River in the Amazon region of Brazil. High levels of mercury are found around small gold mines and processing plants in these regions, and in nearby bodies of water and bottom sediments downstream. Concentrations in fish have reached critical levels. Mercury in the ambient air exceeds permissible concentrations, and clinical symptoms of both chronic and acute mercury poisoning are evident in the exposed populations.

In recent years, global gold production has shifted significantly towards smaller mines. Today, one-quarter of global gold output comes from small mines in developing countries, led by Brazil, the Philippines, Papua New Guinea, Colombia, Ghana, Zimbabwe and Ecuador. Each of these countries produces more than 15 t/y of gold from small and very small operations. Each year, these countries produce approximately 100 t of gold from operations involving mercury, at a usage rate of 4 kg of mercury per 1 kg of gold produced. As a result, these tropical environments are exposed to an additional and highly localised load of some 400-500 t/y of mercury.

While the environmental behaviour of mercury in temperate climates is well-known, little research has considered tropical conditions. Fundamental questions such as the speed and causes of methylisation (the reaction of elemental mercury to form organic compounds) remain unanswered, despite the fact that organic mercury compounds are up to 100 times more toxic and more easily absorbed into the food chain.

Small gold mines in developing countries usually use amalgamation in the gold recovery process, a technology that has long been abandoned in larger units because of its poor recovery, high operating costs and associated occupational and environmental risks. The smaller operations have few technological choices, and continue to depend on outdated and arduous processing technologies. The long-term effects of environmental pollution on the agriculture, fisheries and health of the populations of large river basins will most probably outweigh any economic gains from such small mining.

Agusan River Basin, Philippines

Small-scale gold mining in the Agusan River basin boomed between 1984 and 1988 with the discovery of large near-surface deposits on Mt Diwalwal, Mindanao. At one time the site, some 800 m above the Agusan valley, supported more than 100,000 miners, who organised themselves into groups and co-operatives of between 10 and 1,000 people. Together they produced up to 15 kg of gold per day at the height of the gold rush in 1988. Today, exhaustion of near-surface deposits (up to 150 m depth) and increasing slope failure have limited production, but an ever growing mining population is slowly moving to other sites. Ironically, recent exploration has been greatly facilitated through deforestation, and associated denudation and landslides.

A typical small mining operation consists of only a few tunnels, which follow the vein mineralisation. Placer mining has long ceased for lack of deposits. Milling is done close to the mine, with mercury being added to the makeshift rod mills at a rate of 20 g per kilogram of 10 g/t gold ore. The resulting amalgamated pulp is washed and the amalgam separated, free mercury being recovered from the concentrate by squeezing through a cloth; the mercury is used three or four times before being discarded.

The remaining amalgam is traded by the mine/mill operator and is delivered to crude refineries in nearby towns. Here, it is blow-torched to yield a spongy gold.

Further bullion refining uses nitric acid digestion, and in some places, carbon-in-pulp and electrowinning are used in the gold recovery process.

Mercury emissions occur through all stages of the mineral processing, but are concentrated at the blow-torching of the amalgam, where more than 70% of the mercury losses occur. Minor emissions come from the amalgam-pulp separation washing process, and from volatilisation and spillage during handling. Most atmospheric emissions quickly return to the river system, where the mercury is concentrated in the bottom sediments and is washed downstream during occasional torrential floods. The Agusan River basin receives up to 4,000 mm of rainfall annually, with concentrated downpours of more than 100 mm/h.

Madeira River Basin, Amazon, Brazil

Gold mining in the Amazon is typically centred on river-bottom gravel deposits. Some 650,000 people are estimated to be involved in this type of mining from floating dredges in eight different mining districts throughout the Amazon basin. The gold-bearing gravel is mechanically lifted from the river bottom and is preconcentrated by gravity technology (riffles) before being amalgamated in large cans. The amalgam is burnt off on the dredge or on the river bank, emitting very large quantities (more than 1 kg of mercury per kilogram of gold) into the atmosphere.

The Madeira River basin is located in the southwestern corner of the Amazon. Some 3,000 dredges are currently operating along a 300 km stretch of the Madeira River itself, from Porto Velho up to the Bolivian border. It has been estimated that up to 100 t of mercury have already been discharged into the basin.

After heavy rainfall, the Madeira River overflows into adjacent flood forests. Highest levels of mercury pollution occur in the quiet backwaters of the forest rivers, while the bottom sediments of the Madeira itself are quickly diluted and carried downstream.

3.5.2 Environmental Impact

In both of the cases given above, around 100 t of mercury have been spilled in less than 10 years into a river basin. The effects have been reported in two independent environmental monitoring studies, undertaken around 1988 and using comparable methodologies. While the Brazilian study necessarily covered a wider area, the Philippines study was able to analyse more samples and to identify point sources.

Mercury in the Air: The largest contribution to mercury emissions (40-70%) is generated during the burning-off of the amalgam. In the Philippines, point sources are individual refineries, especially in the city of Tagum, where 21% of the samples of ambient air taken from the streets were found to be above the maximum permissible concentration of 0·015 mg/m^3. Some samples were up to 100 times this level. In the refineries, levels were up to 2,000 times the maximum permissible concentration. In the Madeira River basin, sources are widely distributed, mostly in the open. Therefore, high levels of mercury are only recorded in the immediate vicinity of dredges or blow-torching operations.

Mercury in Water: Most of the atmospheric mercury is quickly deposited in the soil or in bodies of water. Yet, environmental levels of mercury are still relatively low, since the elemental mercury is quickly absorbed on mineral surfaces and is deposited in bottom sediments, methylisation taking place only slowly. With a specific gravity of 13·6, elemental mercury is quickly deposited or is carried away. Nevertheless, approximately 30% of all the fresh water samples taken in the Philippines study were above the maximum permitted concentration of 2 mg/l set by the National Pollution Control Commission and other agencies. There is, however, little scientific evidence that this is an acceptable limit, as natural background levels are substantially lower. Maximum values are reported at 41,000 ppm (4·1%) in mill effluents, and the Agusan River and its tributaries are heavily contaminated. Even 200 km downstream, seawater in the Gulf of Davao – which receives outfall from several mining sites – has mercury levels of up to 200 times those of average ocean water.

Mercury in Sediments: Mean mercury levels at both regions studied are around 50 times the average background level. Maximum levels go up to 12 ppm in the Philippines and 157 ppm in Brazil, equivalent to more than 1,500 times the background value. It is assumed that most of the mercury released into the environment is temporarily locked into the vast quantities of soft sediments and bed loads of the major rivers. It can be remobilised by either a slow digestion and methylisation (through microbial digestion or intake from bottom-feeding fish) or from a rapid remobilisation of soft sediments during flash floods. In both scenarios, which are quite possible, the mercury would find its way into the food chain, and thus create environmental disasters of much larger proportions than are seen currently.

Mercury in Fish: In spite of the potential human health hazard, very few samples of freshwater fish from polluted rivers were analysed for their mercury content during the studies. In both cases, similar results were obtained; mercury in fish is already often above the World Health Organisation's maximum permitted concentration of 0·5 ppm. In the Brazilian example, 17% of fish samples exceeded this limit, while marine seafoods from the Gulf of Davao showed similar effects; 21% of the samples exceeded acceptable levels, with the highest mercury values found being six times the maximum permissible concentration.

Mercury in the Human Population: The Philippines has made an intensive investigation into the health effects of mercury exposure; Brazil has not. Taking workers who had been exposed to mercury for an average of 30 months or who lived within 500 m of some mercury-emitting source, some 72% of those sampled showed clinical signs of its effects. Several cases of acute mercury poisoning and related deaths have been reported from both the Philippines and Brazil.

3.5.3 Response

In both countries, some initial legal attempts have been undertaken to reduce the emissions of mercury into the environment.

In Brazil, small-scale gold mining was declared illegal in March 1988, an unenforceable move that was lifted several months later. Between 1980 and 1985, the use of mercury amalgamation was effectively banned at the Serra Pelada operation, but has been resumed as a result of the problems of policing such a ban. Attempts in 1990 to ban small-scale mining from tribal lands by blasting the supply airstrips met with little success, and as yet no comprehensive national policy has been developed to check environmental pollution from mercury in Brazil.

In the Philippines, public awareness was heightened in 1988 by the publication of reports of mercury-contaminated fish being sold in the provincial capitals of Butuan and Davao. However, little technological change can be expected as long as the majority of these operations is illegal, a situation that has made the provision of technical assistance to the small miners extremely difficult.

In consequence, innovative legislation has been proposed. Under the Minahang Bayan law, still pending approval in the Philippines Congress, certain areas of the country will be set aside as People's Mining Areas with exclusive rights for small operators. In addition, several forms of co-operatives for mine operations are prescribed, and government-run custom milling is to be introduced. A thorough revision of the Mining Code is already under way to protect small miners' interests.

In technical terms, several options have been partially implemented under the supervision of the Inter-agency Task Force on Mercury Surveillance. These include the introduction of settling ponds at the outfall of major mining areas, the zoning of small-scale mills and refineries (into locations away from residential areas), distribution of mercury retorts for recovery of mercury vapour during the refining stage, construction of government-run cyanide leaching plants, and public education campaigns.

In both countries, the question of imposing a ban on mercury has been discussed, and it has been concluded that such a ban would not effectively limit the use of mercury, but would drive more miners into illegality at a time when their co-operation is being sought. Because of the physical impossibility of policing the large areas in question, and of withstanding the pressure from powerful interest groups, neither country has yet been able to tackle the legal questions involved. Progress in the formulation and implementation of adequate laws has been slow.

Despite the attempts in both countries (and, apparently, in similar situations in Papua New Guinea and Surinam) there has been little progress achieved through public education drives in teaching miners and refiners about the safe handling of mercury at all stages. Retorts are hardly ever used in the refineries and, while protective measures such as gloves and masks are usually mandatory, in practice their use is the exception rather than the rule. Finally, the lack of any safe toxic waste disposal systems makes environmental progress extremely unlikely as long as there are substantial amounts of mercury used in remote small mining districts.

What is required is a policy that takes the economic and technological conditions of the small miner into primary consideration while offering feasible non-mercury technology. However, little research and development has been directed to this area. Medium-sized and large operations favour the use of cyanide leaching techniques, but judging from field experience serious doubts exist about the safe operation of such plants in the smaller-scale context.

Few units have full containment facilities that would successfully protect the surroundings from serious cyanide spillage during plant failure, mishandling, accidents or disasters such as earthquakes. In addition, there are reports about misuse and the illegal sale of cyanide solutions for fish poisoning.

Cyanide technology can be handled safely only at large companies, which can provide the appropriate technology and trained staff, and which can be kept under close supervision by the regulatory authorities. In addition, many ores — especially fine-grained sulphide ores — are not easily amenable to cyanide leaching.

It is therefore necessary to look for other, more environmentally benign, processing technologies. In principle, a large array of gold beneficiation methods are available to the small miner. Traditionally, gravity separation has been the most widely used method, combining low machinery investments and operating costs with ease of operation. However, recoveries depend very much on the adaptation of the methods to the local conditions and the gold ore mineralogy.

Sluice boxes have recently received renewed attention and are available in advanced and adaptable designs. They are still the most reliable and cheapest method of pre-concentration. Jigs and tables are useful only in larger operations with higher throughputs. They operate down to a particle size of 0·5 mm, and are now successfully manufactured in developing countries such as Thailand, Malaysia and Indonesia.

Spiral concentrators are successful at even smaller particle sizes, and require less handling. Unfortunately, they are not manufactured in developing countries, and their prices therefore remain quite high. Centrifuges can be considered the state-of-the-art equipment for gold recovery. They have been used widely under field conditions, and have achieved high recovery rates. However, throughputs are significantly higher than those of typical artisanal operations.

Several designs of centrifugal concentrators are now in use. The Knelson concentrator is perhaps the most widely tested and utilised on the market, and is now being built in developing countries, which will bring down the price. The Denver Process Equipment Gold Saver is a low-cost design with proven field applicability, while at the high-tech end of the market, the Mozley Multi-Gravity Separator is now undergoing field testing. Claimed to separate efficiently particle sizes down to 1 micron, its high unit cost has so far limited its application in small-scale mining operations.

3.5.4 Assessment

Small-scale miners will continue to supply a considerable share of global gold production. With deteriorating living conditions in many rural areas and growing population pressure on decreasing land resources, more people will venture into less and less profitable mining districts. They will use whatever processing and recovery technology is available to them; at this time, this is amalgamation. The environmental risks associated with mercury handling cannot be overstated. Several Minamata-type cases are likely to occur.

Technological research into non-mercury and non-cyanide processes is urgently needed, while at the same time the legal framework regarding small-scale mining is in need of overhauling in most developing countries. Governments, research institutions and development agencies need to act together to address the socio-economic conditions that create the situation described in this study, and to prevent further environmental destruction.

3.6 National Environmental Strategies, Poland

Poland depends heavily on its minerals sector for both domestic resources and the generation of foreign exchange. The former centrally planned economy relied heavily on the country's coal, copper and sulphur resources in particular; however, the environmental cost has been high. Poland is now implementing a national environmental policy, aspects of which include the minerals industry. The problem is, however, more than national – it has to be addressed in a regional context. The source material for this study was provided by Professor Stefan Kozlowski of the Instytut Geologiczny, Warsaw.

3.6.1 Introduction

Poland is both the largest of the Eastern European countries and has the most important minerals sector, for it is a significant contributor to world production of three commodities – copper, coal and sulphur. In addition, Poland produces silver, lead, zinc, nickel, cadmium, barite, dolomite, limestone and magnesite.

Both hard coal and brown coal are produced in Poland, the former for both domestic use and for export, and the latter for use in mine-mouth thermal power stations. Poland has been a major hard coal exporter since after the Second World War, supplying global markets. Its copper is traded internationally, while it is also the principal source of non-petroleum sulphur in Europe.

A large proportion of the country's workforce has depended on the minerals industry for its livelihood. At the end of 1986, over 1·1 million people were employed in the minerals and metallurgical industries, representing about 25% of the country's industrial workforce. More recent figures indicate that at the end of the 1980s, nearly 600,000 people made their living from the coal industry, with about 270,000 working underground and a further 160,000 employed on the surface at coal mines.

The 1960s and 1970s witnessed rapid growth in mineral production, and by the late 1970s, Poland was numbered among the top ten mineral producing countries. Mineral production increased steadily until 1978, but after that year it declined rapidly until the economic breakdown of the early 1980s, when martial law was introduced in Poland. Later, mineral production began to rise again, but it never regained the level achieved in 1978.

The change-over from a centrally planned economy to a market economy is still in progress. Of necessity, this has meant a re-assessment of priorities, with uneconomic productive capacity facing closure and major new investment required in the modernisation of mining, smelting, metallurgical and power generation operations. Increased environmental awareness, which came to the fore during the 1980s, is also demanding that the mineral and metallurgical industries should take a more environmentally responsible attitude, reducing the amount of pollution caused and rehabilitating areas damaged by their past operations.

3.6.2 Environmental Impact

For many years, environmental aspects were disregarded by the mining and processing industries. In consequence, the newly created mining districts soon became areas of environmental hazard. At present, 27 environmental hazard districts have been identified in Poland, covering 11% of the country's land area but inhabited by one-third of its population. The worst environmental conditions are found in the Upper Silesian industrial district and in the Legnica-Glogów copper district.

The environmental problems encountered include the degradation of groundwater resources from the vast depression cones caused by mining operations (these cover some 5% of the Polish land area), groundwater pollution by oil-derived products, and the pollution of the Vistula and Ode rivers due to disposal of brines from the hard coal mines. Also of considerable concern is the great volume of mine and industrial dumps; about 170 Mt of industrial waste is produced annually, and in 1988 the total volume of the dumps was estimated at 1,800 Mt. Meanwhile, there has been soil pollution by heavy metals, mainly lead and cadmium, around Poland's iron works and non-ferrous metal plants (especially in the Crakow region, the entire central part of Upper Silesia, and significant parts of the Legnica-Glogów copper district).

At present, environmental hazards create an economic and social barrier for the further functioning of several mining districts. Soil degradation has also become a very serious problem. In 1985, the causes of surface degradation were categorised as:

Ineffective mineral exploitation	38%
Deterioration of the productivity of agricultural ecosystems	20%
Deterioration of the productivity of forest ecosystems	16%
Loss of agricultural, forest and building areas	16%
Reparation of mining damage	8%
Others	2%

It is thus evident that, in Poland, mining is the decisive detrimental factor in degradation of the landscape.

Besides economic losses, surface degradation has very serious social consequences. In Upper Silesia, living conditions have deteriorated rapidly. Today, a high rate of infant mortality, increased general mortality and indications of genetic damage are problems that urgently await solutions.

3.6.3 Response

In the period 1986-1990, a research project was undertaken entitled 'Principles of natural environment management in mining areas'. Under this project, a strategy of lithospheric protection has arisen. This is aimed at the protection of soils against chemical and radioactive pollution, protection of the quality and quantity of both ground and surface water, the utilisation of both the main and associated minerals in a deposit, the protection of the environment around mining and mineral processing sites, and the protection of selected areas that have special natural, tourist or recreational value.

The achievement of these goals will require the effective and economic management of mineral resources, the minimisation of the environmental impact of thermal power generation, tight control of the quality and quantity of chemical production, and of toxic chemicals in particular, and the special protection of surface, marine and ground waters.

The strategy of lithospheric protection has both global and regional dimensions. Before the agreement of international programmes, local activities should be initiated in every country. In this respect, Poland's needs are particularly urgent as its industrial pollution is believed to one of the worst in the world.

The main and most urgent steps to be taken towards lithospheric protection are the limitation of the amount of toxic waste introduced into the soil and groundwater, the promotion of low-waste and waste-free technologies, the clarification of legal problems with regard to mineral resources and groundwater protection, the establishment of control policy mechanisms, and the setting up of precise executive regulations with respect to soil protection as well as effective control mechanisms.

There is a vital need to create a monitoring system for the environment, covering such areas as ground water quality — such a system is now being introduced by the State Geological Institute — and geochemical changes, mainly in the subsurface zone, to be presented as a geochemical map of Poland.

A problem in itself is the protection of mineral resources, which is covered by the Protection of the Lithosphere programme. This is now in progress at the State Geological Institute, and is to be completed by the year 2000. The programme is divided into seven subject groups:

1. The evaluation and mitigation of hazards, and management of lithospheric resources
2. Protection of the earth's surface against chemical and radioactive pollution
3. Protection of the lithosphere based on geophysical results
4. Protection of groundwater
5. Protection of mineral deposits
6. Utilisation of minerals in environmental protection
7. Protection of selected areas of special natural, tourist and recreational significance.

This research is parallelled by changes in the old legal system, which reflected the central planning approach and Marxist philosophy. An amendment to the Geological and Mining Acts was passed in the Polish Parliament, introducing a number of new regulations adapted to a market economy. The most important steps are the definition of the right of property of mineral resources, the setting up of terms for obtaining exploration and mining licences (including foreign companies), the establishment of concession and mining fees, the definition of the role and responsibilities of the Polish Geological Survey, the foundation of the Geological Council, and the definition of the role and responsibilities of the Higher Mining Office.

All the executive regulations of the amended Geological and Mining Acts are to be issued by the end of 1991. As these legal measures were produced after consultation with the World Bank, Poland will be the first of the former communist countries to possess mineral legislation that is compatible with the requirements of the European Community.

3.6.4 Assessment

There is no doubt that the minerals sector is in an invidious position. Having formed a firm foundation for both domestic industry and for hard-currency exports for many years, it is now being held responsible for a high proportion of the country's environmental problems.

The problems are by no means confined to specific small areas of the country, but are widespread. In addition, the south of the country in particular is a net importer of airborne pollution from neighbouring countries such as Czechoslovakia and what was formerly East Germany. One of the most visible signs of environmental damage is forest denudation; if drastic action is not taken to reduce atmospheric emissions, it is estimated that by the year 2000 one half of Poland's forests (covering about 4 Mha) will have been lost.

Environmental protection is just one of the problems facing Poland during its transition to a market economy. The publication of a national environmental policy document reflects the seriousness with which the country regards its new-found responsibilities in this area. However, its chances of unilateral success are limited; the problems that it has inherited from the past 40 years are too great for Poland to handle alone.

Poland's environmental situation, in which its minerals industry is a key player, has to be seen, therefore, in a European context. While great efforts have to be made to improve the situation within the country, a multilateral approach is urgently required to ensure that Poland's task is made easier, and that benefits will be seen both within the country and beyond its borders.

Participants at the Mining and Environment Round Table, Berlin 1991.